LAND USE AND
WATER RESOURCES

LAND USE AND
WATER RESOURCES

IN TEMPERATE AND
TROPICAL CLIMATES

H. C. Pereira F.R.S.

CAMBRIDGE

at the University Press

1973

Published by the Syndics of the Cambridge University Press
Bentley House, 200 Euston Road, London NW1 2DB
American Branch: 32 East 57th Street, New York, NY 10022

© Cambridge University Press 1973

Library of Congress Catalogue Card Number: 72–85437

ISBNS: 0 521 08677 9 (hard covers)
       0 521 09750 9 (paperback)

Printed in Great Britain by
William Clowes & Sons, Limited
London, Beccles and Colchester

Objective

This is a summary in plain language of the information at present available to guide decisions on policy for land-use management in watersheds. Methods available for measuring the hydrological effects of land-use changes are described. Results both of research and of recorded experience are discussed in their relevance to temperate and to tropical environments.

Contents

Introduction   *xi*

Conversion factors   *xiv*

1 The world's water resources and the growing demand   **1**

The hydrological cycle   I
The quantities of water in circulation   4
    dew 5
The development of water resources   6
The dawn of public awareness   7
    population growth 7; water demand 9
Groundwater storage and development   II
    *radio-active tracers for groundwater 12; underground
    reservoirs 13; estuary storage 14*
Desalination for domestic supplies   15
    solar energy for distillation 16; atomic energy 17
Desalination for irrigation   17
    oil-fuelled processes 18; basis for comparison of costs 20
Rain-making   20
Summary of resources   21

2 Development of a watershed discipline   **22**

The historical choice of discipline or disaster   22
Organisation for mutual protection on a watershed area   23
    floods 23; soil conservation 23; economic progress 24
Protection of streamsource areas   25
    critical areas of erosion hazard 26
Multiple land-use planning to include recreation opportunities   29
    *urban concentrations 29; public access to protected catchments 30;
    management of wildlife populations 30; management for
    preservation of scenery 31*
Management for maximum water output   32
    *conservation v. collection 34; water loss by evaporation from
    reservoirs 35; community organisation by watershed boundaries 36*

3 The achievement of hydrological information   **38**

The first stage in research and development of water resources   38
    *the variable input 38; the dividends of foresight 38; priorities in
    developing countries 39; routines and research 41*

The sampling of rain and snow   41
    the catch of raingauges 41; measurements by radar 44;
    assessment of snowfall 44
Streamflow measurements   45
    routines and problems 45; rating by radio-isotopes 46; weirs and
    flumes 46; analytical studies of streamflow records 49
Groundwater measurements   50
    conventional means of exploration 50; flow and direction
    from single boreholes using radio-active isotopes 51;
    interpretation of results 52
Measurement of changes in soil-moisture storage   53
Evaporation and transpiration   55
    quantitative estimates 55; evaporation pans 56; evaporation from
    vegetation 58; potential evapotranspiration 63; advection 67
Recording of land-use changes   69

4   Recorded experience of the effects
of forests on watersheds   **72**
The effects of forests on weather   72
    forests and mists 72; snow-trapping by forests 73;
    forests and rainfall 74
The effects of fire on watersheds   75
Beneficial use of water by forests   76
Forest plantations and streamflow in warm climates   78
Afforestation and water supplies in cool climates   81
Reafforestation of eroded watersheds   81

5   Research on forested watersheds   **84**
Streamflow comparisons   84
    clear-felling experiments 85; multiple-valley experiments with
    forest plantations 86; summary of comparisons of streamflow 87
Balancing the budgets for both water and energy   89
    seasonal soil-moisture storage changes 91; seasonal groundwater
    storage changes 91; leakage of watersheds 92
Measurement of incoming energy   93
    orientation of catchments 97
Studies of special forest components of the water balance   98
    interception of rainfall 98; interception and condensation of cloud,
    fog and mist 102; interception of snowfall 103; effects of forests
    on snowmelt 104; effects of streambank vegetation 105;
    control of stormflow 107; integration of results by computer 109;
    urgent needs for experimentation 113

6   Watershed experiments in tropical forests   **115**
An outdoor laboratory in high-altitude tropics   115
    a research opportunity 117; watershed experiments in East Africa
    117; can productive tree plantations safely replace
    natural forest? 118

Replacement of tall rain forest by tea plantations   119
    results: water use 122; checks for leakage 122; control of
    stormflow 125
A simplified approach: studies on plots of trees   127
    a fifteen-year test of a soil-moisture budget 128; a soil-moisture
    budget for tea irrigation 129
Bamboo forest or pine plantations in a mountain watershed   131
    watershed comparisons 135; measurements 137; results 137;
    check for leakage 138
Application of energy–budget analysis to the records of watersheds
   in the USA   138
The future role of watershed experiments   140

7 *The effects of grazing animals on watersheds*   142
Grassland as a land use   142
Grasslands well supplied with water   143
Effect of grazing on marshlands   144
Snow-trapping on cold rangelands   145
Grazing in forested watersheds   145
Grazing by forest wildlife   147
Grasslands having long dry seasons   148
Watershed experiments in range improvement   149
    a study in East Africa 149; regeneration of tropical grasslands 153;
    evidence from Australia 154; water use by improved rangeland 158
Semi-arid grassland   159
    range management for watershed control 160; sediment flow 161;
    flood-spreading on rangelands 163; heat reflection from dry
    grassland 165
Summary of effects of grazing on watersheds   166

8 *The effects of croplands on water resources*   167
Watershed behaviour   167
    surface soil management 167
Cropping in climates of seasonal drought   170
    soil-conservation effects in the USA 170; practical examples of
    hydrological improvements 172; experiments in the USSR 175;
    summer rainfall and winter drought 175; tropical problems 176
Cropping in climates of water excess   180
Semi-arid croplands   181
    a mechanism for survival 181; the penalties of misuse 182
Protection of soil and water on arable lands   183

9 *The roles of irrigation and drainage in water*
**   *resources*   184**
Groundwater and salinity   184
    a practical example 186

Effects on major rivers 186
    the River Volga 186; the Murray River 187
Effects of over-pumping of groundwater 190
Efficient use of water for irrigation 191
Biological hazards of irrigation 191
Water harvesting for supplementary irrigation 192
Drainage of marshlands 194
    effects of swamp drainage on water quality 198; effects of swamp drainage on flood control 198; ecological effects of marsh drainage 199

10 Problems and priorities 201

In countries of advanced technology 201
    watershed effects of urban and highway development 201; industrial pollution 202; pollution from modern agriculture 206; watershed control and countryside amenity 208
In developing countries 208
    urban development and pollution 208; highway drainage 209; control of land use in streamsource watersheds 209; international aid 211; economic development 211; the spread of subsistence agriculture 212

Recommended reading 215

References to literature 217

Subject index 241

Introduction

The following pages are intended to be a convenient form of background reading for both the lay and professional members of the many public bodies which carry responsibility for water resources and the management of watersheds. They do not form a research treatise and are not intended as a textbook.

Water resources are belatedly receiving a due measure of research attention in most countries of advanced technology, but the current fashionable enthusiasm for ecological protection frequently confuses the issues. There is a real need for better informed public opinion on matters which already affect our daily lives and may well dominate the lives of our grandchildren. In temperate climates the general public of the Western World are newly awakening to the realisation that our technical progress already outruns our ability to protect vital natural resources. Fresh air, fresh water, good agricultural land and areas of great natural beauty are belatedly recognised as being in need of active protection on a national scale.

On the international scale the political importance of these issues has increased rapidly over recent years, from the platitudinous Biosphere Conference in Paris in 1968 to the sharply controversial Conference on the Human Environment at Stockholm in 1972. Political enthusiasm is indeed necessary to generate legislation and to raise funds, but decisions for action must depend on knowledge of causes and effects. Knowledge of the ways in which our management of land can conserve or destroy our water resources is won by the patient recording of field measurements, often employing substantial resources over many years. The Stockholm Conference has emphasised the urgency for decisions on the protection of natural resources: many decisions on land use must be made on the fragmentary evidence which we now have. The interpretation of existing evidence on the hydrological effects of the management of watersheds is therefore an increasing preoccupation of many water resource authorities.

Public concern is often channelled through Committees of Enquiry. These committees are deluged with opinions stemming from every kind of personal and communal interest, quoting and mis-quoting figures and facts both in and out of context. Much of the technical evidence for developments which could threaten our environment is assembled by civil engineers who call on organised specialist services from laboratories dealing with such subjects as air and water pollution. The engineering aspects are therefore mentioned only briefly and there are no complex details of water treatment chemistry. Civil engineers, however, have to cover a formidable range of subjects and to make early and practical decisions. The many combinations of climate, rocks, soils, vegetation and agricultural development often hinder recognition and prediction of the effects on water resources when changes are made in the use of land. Such evidence is scattered among the literature of meteorology, geology, hydrology, soil and plant sciences, agriculture and forestry.

Direct experiments using the combined techniques of these many environmental sciences have now achieved more conclusive results than the confusion of literature would suggest. The more important of these results are brought together in the following chapters.

Technical terms are used sparingly but references are given throughout to specialised source material in which details can be found, so that this book may, it is hoped, prove helpful alike to teachers and students.

The metric system is used in the text, with a few exceptions where direct quotations are made from reports in British units. Quantities are given in both units where this may help some readers to follow the argument more readily. A conversion table is given on page xiv.

A literature survey has been made over a wide field but a full catalogue of references would be both unwieldy and unnecessary: the citations are selected as far as possible from land-use studies known to me personally. In order to keep such a survey to a tolerable length, only examples can be quoted, but by following up the references a rich field of further work will be encountered.

Individual acknowledgement to the colleagues in many lands who have shared in experiments over the years, or who have given freely of time and effort to visit and to discuss their own studies with me, would be impossible over so wide a field. My thanks are none-the-less sincere because they must, perforce, be collective. Acknowledgements of the sources of information are made throughout the text and illustrations.

Financial support from the Overseas Development Administration of the UK made possible the watershed experiments in East Africa described below, and my thanks are due to the staff of the Food and

Agriculture Organisation of the UN for operational support on many of the travels in search of information for the International Hydrological Decade.

I thank Dr J. S. G. McCulloch and Dr J. R. Blackie for reading the manuscript and for their valuable suggestions.

Finally my thanks are made both to the ODA and to Sir Joseph Hutchinson CMG, FRS, for three months of hospitality at the Cambridge School of Agriculture which made possible the beginning of this book.

East Malling   H. C. PEREIRA
1972

Conversion factors

	To metric units	From metric units
Length	1 inch = 25·4 mm	1 mm = 0·0394 in
	1 foot = 0·3048 m	1 m = 3·281 ft
	1 yard = 0·9144 m	1 km = 0·621 mile
	1 chain = 20·12 m	
	1 mile = 1·609 km	
Area	1 in² = 645·2 mm²	1 m² = 10·764 ft²
	1 ft² = 0·0929 m²	= 1·196 yd²
	1 yd² = 0·8361 m²	1 ha = 2·471 acre
	1 acre = 4047 m²	1 km² = 247·1 acre
	= 0·4047 ha	= 0·3861 mile²
	1 mile² = 259·0 ha	
	= 2·59 km²	
Volume	1 ft³ = 0·02832 m³	1 m³ = 35·315 ft³
	1 yd³ = 0·7645 m³	= 1·3080 yd³
	1 gal(UK) = 4·546 litres	1 litre = 0·220 gal(UK)
	= 0·004546 m³	= 0·264 gal(US)
	1 gal(US) = 3·785 litres	
	= 0·003785 m³	
	1 ft³ = 6·23 gal(UK)	
	= 7·48 gal(US)	
	= 28·32 litres	
	1 mile³ = 4·166 km³	1 km³ = 0·2399 mile³
	1 acre-ft = 271,400 gal(UK)	100 mm rainfall or irrigation
	= 1613 yd³	approximately
	= 1233·5 m³	= 4 in
	1 million gal(UK)	= 530 yd³/acre
	= 3·7 acre-ft	
	= 4564 m³	

Flow   1 cusec (ft³/s) = 0·02832 m³/s (cumec)
        1 million gal(UK)/day
                = 3·7 acre-ft/day = 4564 m³/d (cmd)
                = 1·867 cusec   = 0·0531 m³/s
        1 cusec for 24 hr = 1·983 acre-ft = 2447 m³

| Power | 1 horsepower = 745·7 W | 1 kW = 1·341 hp |

1

The world's water resources and the growing demand

The hydrological cycle

About four-fifths of the world's surface is covered by the oceans: the surface of this vast area of salt water, where radiant energy is absorbed from the sun and fresh water is evaporated into the atmosphere, is the source of the supplies which maintain our lives (Fig. 1). The term 'hydrological cycle' describes the world's circulation of fresh water as it evaporates from the sea into the atmosphere, is transported by the wind until it condenses as water droplets into clouds and is precipitated as rain, snow or hail. We do not know how much falls back directly into the sea, but it is estimated at up to three-quarters of the total. The sun's radiation drives the circulation of the fresh-water cycle, firstly by providing energy for evaporation from the surface of the oceans and secondly by differential heating of the air masses, which provides the energy for transportation of vast quantities of water vapour across the surface of the globe.

Water falling on the land surface may run off directly into streams to drain by way of rivers and lakes back into the sea, but the most important pathway for the sustenance of man is that of infiltration into the soil. From the soil vegetation is supplied; the surplus, seeping underground to springs, maintains the steady flow of rivers. Soil depths range from a few centimetres to 15 metres or more, being characteristically more shallow in the temperate zone and deeper in the tropics, as a result of the contrasting geological histories of these zones during the Ice Ages. Plant roots explore more deeply in areas with long dry seasons and have been traced to 15 metres in tropical soils. Part of the water stored by the soil within these root ranges is available to plant roots and is transported upwards through the plants to be evaporated (transpired) from the foliage. Transpiration is an important process in the hydrological cycle and has received much intensive study. It is now known that the rates of

I

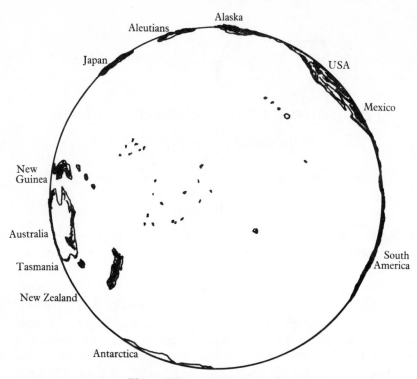

Fig. 1. The evaporation source.
About four-fifths of the earth's surface is covered by water. Solar energy absorbed by this vast surface provides the energy to operate the hydrological cycle.

transpiration are controlled by the plants only at extreme stages in their development but that, for the most part, the rates are determined by the heat energy available for evaporation, the ability of the air to carry away the water vapour and the ability of the soil to supply water to the roots. The explanation of this physical basis of plant water use has been achieved only in the last twenty years, as the result of long and patient research. Some water evaporates directly from soil and plant surfaces and from the water surfaces of lakes and rivers. A little condenses from the atmosphere onto the earth's surfaces as dew on cold nights. The hydrological cycle is completed by the return of water to the sea by river-flow and by deep underground seepage. A little returns overhead by evaporation from the land surface and direct precipitation into the sea (Fig. 2).

The proportion of rainfall which follows these different paths is determined by the main geographical factors of climate, topography and

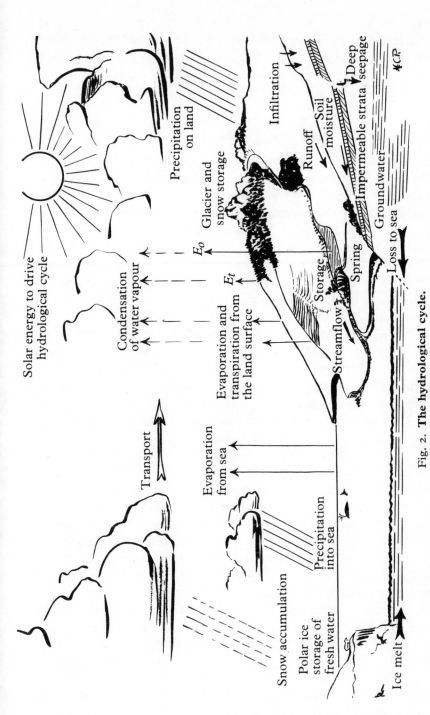

Fig. 2. **The hydrological cycle.**

About 90% of the precipitation over the land is derived from the sea and only 10% comes from evaporation and transpiration from the land surface.

vegetation. Climate, in which the dominant factors are the supply of water and also the energy to evaporate it, is beyond the control of man, although we can ameliorate its local effects by irrigation, shade and shelter. Limited local successes in the initiation of rainfall and in the prevention of hail are discussed later in this chapter.

The quantities of water in circulation

Several estimates have been made of the world's total water quantities, and since parts of the estimate can only be guesswork, the results vary substantially. There is general agreement that about 97% of the total water is held by the oceans, although since our knowledge of the topography of the sea bed is far from complete the total volume is not accurately known. Changes in sea level can be measured only by reference to the land surface, but earthquakes and volcanic activity are only the more noticeable evidence of the complex history of movement of the earth's crust. No exact figures can therefore be given, and an exact quantitative estimate of the world's water budget is not yet possible. Approximately $3\frac{1}{2}$% of salts in solution in this vast body of water represents the difference between life and death to land animals and plants. Three-quarters of the world's fresh-water supply is held frozen in the polar ice-caps of Greenland and Antarctica. These great storage reservoirs of fresh water do not remain constant. Greenland is slightly losing water, the sea-ice and the glaciers having retreated in the last century, and radio-carbon dating suggests that enough ice and snow has melted from the ice-caps in the past 18,000 years to raise the sea level substantially. Meanwhile the Antarctic continent gains about 600 cubic kilometres per annum, but although the rate now appears to be declining, the Antarctic's gains more than match Greenland's losses. The effects of changes in these two major reservoirs probably far exceed any effects of man on the world's fresh-water resources (Orvig 1970).

The fresh water in lakes and rivers and in known underground fresh-water resources available to man is less than 0·05% of the world's total water content. It is, nevertheless, estimated to be more than 500,000 cubic kilometres (120,000 cubic miles). If better distributed this would be ample for all of man's foreseeable requirements. In addition there are large supplies of underground water too contaminated with salts to be immediately useful, but much of it is classed as 'brackish' (i.e. having no more than 1% of dissolved salts): this has limited uses for irrigation, for which it must be diluted by water of better quality, before it can be counted as an additional resource.

4

The annual total of river discharge into the oceans has been generally accepted to be equivalent to about 10 cm change in the sea level. As a flow this amounts to 924 cubic metres per second (or 29,100 km³ p.a.), although our totals of measured river flows do not reach this figure. Only seventy of the world's rivers make much difference to this total, but we have inadequate data for some of the great rivers of Asia and only rough estimates for the flow of the Amazon, the world's largest river. Floods are very difficult to measure and there are also estimated to be substantial but unmeasured losses of fresh water into the sea by underground seepage along some 370,000 km of coastline (Nace 1970).

The total water content of the atmosphere, mainly in the first 4·5 km (3 miles), is only about 25 mm or 1 inch but its distribution is highly erratic. Atmospheric circulation is due mainly to the temperature differences between the Equator and the Poles but it is subject to very large-scale random fluctuations (Sutcliffe 1956). Recent cloud photographs from space-craft outside the atmosphere offer new and powerful methods of study of these patterns and fluctuations, but as yet the evidence and methods of analysis are still being developed.

When the wind systems cause air masses to converge they are forced upwards into colder conditions and precipitation of rain, hail or snow occurs. Only a small part of this water falls where it can be used by man. One-third to one-quarter falls onto the 140 million km² of land surface, but FAO estimates show only 30% of this area to be suitable for cultivation. Evaporation and transpiration from the land surface make a surprisingly small contribution to local rainfall. Rainfall is clearly not determined directly by the local water content of the atmosphere since coastal deserts such as those of the Red Sea persist although the air above them is often moist. Studies ranging in scale from a valley to a continent (Holtzman 1937; Benton, Blackburn and Snead 1950; Budyko 1958) all agree that over large land masses 90% of the precipitation is of maritime origin and only about 10% is derived from vegetation and from fresh-water surfaces. Although evaporation from the land surface can thus make only minor contributions to the annual totals of rainfall, it can still have a very strong influence on local weather, especially on the occurrence of mists or fogs which are important for visibility over roads and airports.

Dew. A little water condenses from the atmosphere on clear calm nights as dew and this has been suggested as an important source of water for plants. Exaggerated claims, sometimes based on the dew yielded by iron roofs and other surfaces affording rapid heat exchange, have been made

5

for the dewfall over a landscape, but these are not based on physical reasoning. Monteith (1957) has shown by accurate lysimeter measurements that the *annual* dewfall over the British Isles is only from 2 to 5 mm. Dew formation is heavier in tropical climates, where it can be locally important if harvested rapidly, before evaporation, as by grazing animals at dawn. The author has seen, in Tanzania, herdsmen in dry grazing lands collecting their drinking water at dawn by sweeping the dew from tall grasses. The daily dew amounts are small, however, in comparison with the potential daily evaporation.

The development of water resources

Historically, man's need to improve natural resources of water developed first in the warmer latitudes where evaporation exceeds rainfall for several months of every year. Structures for the control, storage and distribution of runoff water were developed by the earliest civilisations: in Egypt, in Babylon, in India and in China. Dam building is thought to have begun on the Nile, where there are, near Memphis, the remains of a masonry wall built across the river by King Menes in 4000 BC. The Romans built dams for water storage so well that some of their structures, such as that illustrated in Plate 1 from Jordan, are usable today (Reifenberg 1955). In the cooler northern latitudes river flows were adequate both to provide water and to remove wastes from towns along their banks.

Population growth, compounded by the great increase in the consumption of water per head by a modern technological society, has caused a

Plate 1. A Roman reservoir still holds water in Jordan.
(Photograph by Department of Antiquities, Jerusalem. Reifenberg 1955.)

6

vast increase in the rate of man's dam-building efforts in the past fifty years. Three great new lakes for storage (Lake Volta is over 300 km long above the Akosombo Dam in Ghana, Lake Kariba 280 km long on the Kariba Dam between Rhodesia and Zambia, Lake Nasser over 300 km long on the Aswan High Dam in Egypt) have all been completed in the last ten years and two of them are still filling. Almost none of the Nile's annual flow of 92 km^3 now reaches the sea: the new reservoir alone can hold 130 km^3. The chains of dams controlling the great Russian rivers, the Dnieper and the Dniester, the Don and the Volga, the great Boulder Dam creating Lake Mead in the USA and the intensive cascade of dams in the Tennessee Valley, the seventeen dams of the Snowy Mountains River system in Australia and the great systems of dams and barrages controlling the Indus and the Ganges, all reflect, by a fine series of engineering achievements, the incessantly increasing demand for two of the basic requirements of modern man, water and electric power. They also make possible major increases in irrigation by which water is supplied where there is sunshine enough for heavy crops.

These impressive developments merely emphasise the scale of the problems facing mankind in the immediate future: the term immediate is no exaggeration when the time-span of dam building, some 6000 years, is compared with the time-span of only three decades to the end of this century, with which most current water-supply plans and population forecasts are urgently concerned.

The dawn of public awareness

There is growing recognition among professional observers that a crisis in human ecology is approaching. The relation of people to their environment is becoming more critically important, for very different reasons, in some of the most technologically advanced countries and also in many of the less developed countries in the tropical and sub-tropical world. In the countries of higher latitude and more advanced technology the impact of television, radio and press, quoting and misquoting a spate of recent papers delivered to international public meetings, has begun to cause a public unease which is already having strong political effects. These are further discussed in Chapter 10. Here we are concerned with the basic causes.

Population growth. The essential problem is that the growth rate of scientific and technological knowledge in the last century has far exceeded the rate at which we have been able to change human social

7

patterns in order to accept and use the new powers which have been won for us. We have used every new advance to reduce human death rates, but have given only tardy and reluctant attention to any corresponding control of birth rates. The facts are startling. World population appears, from the evidence available, to have increased very slowly over several thousand years of early history, at less than one per thousand per annum,

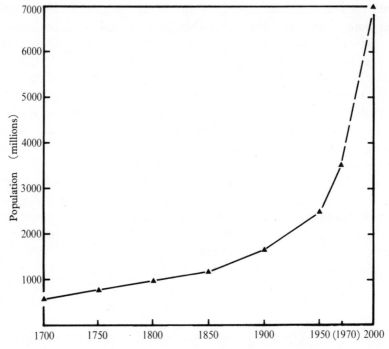

Fig. 3. **World population estimate (millions).**

Many social, economic and biological factors may intervene to alter the shape of the curve over the next three decades but if present rates of increase are sustained world population would exceed 7000 million by 2000 AD.

rising slowly as agricultural techniques improved but still checked by famine and pestilence.

In the countries affected by the Industrial Revolution there was a sharp increase to ten or more per thousand per annum, which demographers ascribe to a fall in the death rate; this was soon followed by a fall in the birth rate so that the technically advanced countries have a range from 3 to 10 per thousand per annum with only slow rates of change (US National Academy of Sciences 1969).

These countries, however, account for less than a third of the human

8

population. In the developing countries the impact of science was felt later, with little change before the Second World War. Since 1950 the effects of DDT on the elimination of malaria, effects of antibiotics on child and maternal mortality, and effects of the anti-sera on major epidemic diseases, have removed some of the more severe biological checks. This was reinforced by the efforts of pioneer administrators and engineers (including the very effective field staffs of the Colonial Services), by the improvements to public water supplies and by a better distribution of foodstuffs through the construction of roads and railways. The international organisation of aid and the shipment of large quantities of grain to counter major national disasters of flood and famine have become effective; all these excellent developments have halved the death rate between 1920 and 1965 so that growth rates now range from 20 to 40 per thousand per annum and are thus doubling in 18 to 35 years (Fig. 3). The general estimate currently accepted for the world population increase is 21 per thousand per annum, doubling in 33 years. The world population in 1970 was estimated by the United Nations to be 3500 million and at present rates of increase it will double in the next three decades, i.e. by 2000 AD it will be 7000 million, in the lifetime of many currently concerned in this debate. To those who take either a fatalistic attitude or one of passive hope, it is necessary to point out that had this rate continued from the time of Christ, there would long ago have been no further standing room for humanity on the dry land surface of the earth. Present rates of increase are therefore critical factors in the planning of land and water resources.

Water demand. We are concerned here, however, not with human numbers but with the water supply to sustain them and with the optimum management of land surfaces on which it falls and from which it must be collected. The World Congress on 'Water for Peace', convened in Washington in 1967, focused attention on mankind's potentially critical water-supply position. In the most developed countries, where public authorities plan engineering and hydrological surveys to forecast water requirements for many years ahead, there is now widespread concern about the rapidly converging trends of supply and demand. In the well-watered temperate zones the past history of plentiful supply has led to the widespread use of rivers to dilute sewage and industrial effluent and to transport these wastes to the sea. It has been officially reported that more than 80% of the total fresh water used in the USA is for waste disposal (US Senate Committee 1960). The pollution aspects of this problem have become urgent in many industrialised countries and are

9

discussed further in Chapter 10: the effect on the supply situation is already serious.

Domestic water use varies with the climate and with the stage of sophistication of the urban community. In Western Europe it varies at present between 100 and 250 litres per person per day, while the higher official forecasts, such as that for Southern England (Water Resources Board 1966) predict 300 litres per day for 1980 and 360 litres for 2000 AD. Current domestic use in the USA already averages some 400 litres per day and official forecasts predict 660 litres per day by 1980 and 1000 litres per day by 2000 AD (US Senate Cttee 1960; Murray & Reeves 1972).

An important change is developing in the pattern of demand, as industry calls for an ever-increasing proportion now amounting to one-half of the total supply. In 1966 a survey, by the Economic Commission for Europe, of twenty-nine countries, estimated total water use at 330 m^3 per annum or 900 litres per day per head of population. Of this only 14% was municipal and domestic, 38% was agricultural and 48% was industrial. The 1970 estimates for the USA are 7%, 36% and 57%. In Britain the rising consumption by industry is paralleled by a major new demand by agriculture for irrigation. Although the islands have a characteristic excess of rainfall over evaporation, research at Rothamsted, Britain's largest and oldest agricultural research centre, has shown that in five years out of ten there is a deficiency of rainfall south of a line from the Humber to the Severn, such that irrigation can raise agricultural yields (Penman 1950a, 1962, 1970). In the south-eastern corner, Suffolk, Essex and Kent have water deficits during the growing seasons in nine years out of ten (Smith 1967). Irrigation research has been taken up successfully in British farming practice but water supplies are severely limiting and present sources are almost fully committed.

The combination of rising population, rising water consumption per head, and rising volume of domestic and industrial wastes for disposal, is outstripping the geographical resources of the environments of major cities. As cities grow, civil authorities must range even further afield in search of water supplies; Los Angeles is now drawing water from 200 miles to the north and 250 miles to the east, while New York taps water from 140 miles upstream. The problems are reaching critical levels of decision at national government scale in many industrial countries including Britain, France, Germany, the USA, the USSR and Japan. Purification plants for the recovery of part of the vast quantity of water used to operate sewage transport systems have become a matter of national as well as civic concern because of their high capital cost and substantial operating costs. President Johnson, in his 1966 Report to

10

Congress on the State of the Nation, announced federal financial support for action to clean the nation's rivers, and in January 1970 President Nixon announced to Congress an 800 million dollar programme for the reduction of water pollution. In Britain a national effort is in progress, led by the Department of the Environment (Chapter 10).

The Thames, in dry weather, falls to a level at which the input of over 100 million gallons per day of treated sewage effluent receives only a 3–1 dilution (Bowen 1962). In addition to recovering more water and polluting less of our environment with wastes, the growing water demand must be met by development of ever more seasonal and over-year storage, steadily reducing the annual loss of fresh water to the sea. Sites with natural advantages for dam building are scarce and where developed farming country is to be flooded the costs and loss to the community may become prohibitive. Storage underground and storage behind barrages in estuaries are two major alternatives.

Groundwater storage and development

The world-wide scale on which groundwater is used, in a somewhat haphazard manner, from boreholes and wells has not yet been matched by an adequate development of research and exploration of the underground water resources. The geology and hydrology of underground water is still very incomplete even in highly developed countries, while in the slightly populated and semi-arid regions of the world underground aquifers are largely unexplored. Historically, this has been because methods of study were costly and laborious. Geological mapping cannot rely only on outcrops and must in large part depend on drilling of boreholes supplemented by surface geophysical methods such as plotting of the earth's local magnetic field, the speed of propagation of shock waves from small explosions, local variations in the force of gravity, and direct measurements of electrical resistance. Where water is located the extent of the water-bearing strata or aquifers is estimated by experimental pumping with simultaneous monitoring of water levels in other boreholes. Since rock formations are often complex and many of the water-bearing materials are sands and gravels in the form of irregular lenses, there has to be a great deal of approximate interpolation between points of measurements. Although far too little direct study of underground water resources has been undertaken in the past, it is fortunate that much exploratory drilling has been done in searches for oil, coal and mineral ores. Powerful new methods of integrating many observations of the fluctuations in groundwater level are now provided both by electrical analogue models and by simulation studies on the digital computer (Cole 1969; Tyson and Weber 1964).

Radio-active tracers for groundwater. New tools for groundwater study, based on radio-active tracers, have been developed into field techniques within the past five years. These are described in Chapter 3. Fallout of 'tritium' (radio-active hydrogen) from atomic explosions in

Plate 2. Laboratory measurement of natural tritium in water samples.

The tritium measurements are valuable but time-consuming. First the small fraction of the water sample in which the hydrogen is radio-active must be concentrated by electrolysis. It is then reduced in an electric furnace to hydrogen gas, which is collected on absorbent material cooled by liquid air. The research worker in the photograph is refilling the cooling flask. The hydrogen is then added to pure ethylene and converted to ethane gas, which is circulated through electronic counters to measure the radio-activity.

the atmosphere has also provided peak concentrations at known dates which act as markers in studies of the slow movement of water underground. Radio-activity measurements also help to distinguish 'fossil' bodies of water which have been trapped from long past periods of abundance and are no longer recharged by fresh supplies. The vast body of fresh groundwater underlying the Algerian Sahara is calculated to take 35,000 years to seep down from the Atlas Mountains. The French engineers discovered it by drilling in 1856 and the extractions at the oases are already of the same order as the probable supply (Furon 1963). Such water can of course be 'mined' but the supply will fail when the reserves are exhausted. The chemical preparation and radiation counting of the samples in the laboratory are slow and elaborate processes, especially with tritium (radio-active hydrogen), which must first be extracted by complex apparatus (Plate 2).

Underground reservoirs. Aquifers may be used as underground reservoirs, into which water is led in times of plenty to be withdrawn by pumping when needed; this is an important modern development of hydrogeology and water engineering. Thorough geological and hydrological explorations are necessary to give assurance of freedom from contamination and from excessive leakage. The advantages are substantial: the water is stored without evaporation loss and without the vast capital costs, maintenance charges and slow destruction by sedimentation which are the fate of many surface reservoirs. This method has been in active use for a decade by the California Water Board (1957). Clean water must be used in most cases to avoid blocking the porous infiltration surface but crude forms of biological filter may be used. Industrial effluent in the form of water used for cleaning of vegetables for canning has been sprayed onto scrub thicket in New Jersey for more than fifteen years, at rates of up to 200 inches per annum, without such blockage. Large-scale experiments both in Israel (Amramy 1967) and in California (Stoyer 1967) have successfully demonstrated the recovery of sewage water, after treatment by successive anaerobic and aerobic stages of storage, and subsequent infiltration into groundwater aquifers. By pumping from points distant from the recharge area, dilution of the treated water results and a long underground storage interval between recharge and re-use further reduces the populations of bacteria. Natural recharge from annual rain may be adequate to refill aquifers which are supplied by a large enough catchment area: the Thames Conservancy has recently developed the system of pumping from capacious chalk aquifers in the Kennet Valley to increase the flow of this tributary of the Thames at times of water shortage. The chalk is recharged by winter rains.

Israel's total water resources are estimated to be 80% underground, with the Jordan River, storm runoff and reclaimed sewage water together making up the remaining 20%. The policy adopted aims to direct half of the surface waters into underground storage by various techniques of recharging of aquifers (Wiener 1967).

A major problem of groundwater storage is the salinity of many rock strata, by which water reaching them becomes unfit for use. This is a critically important problem in arid areas such as Kuwait and in semi-arid areas such as Israel. Rainwater percolating into the hills around the Sea of Galilee passes through saline strata and emerges as strongly saline springs along the coast of Capernaum and in the bottom of the lake. The river Jordan flows into the lake carrying only 30 parts per million (ppm) of dissolved salts and leaves the lake bearing 300 ppm. An elaborate sampling study was carried out, in which 800 samples were taken in one day, in a pattern of concentric circles covering the whole lake bed. The salt-water springs were found to be near to the intake of the main irrigation pipeline and were estimated to be adding 180,000 tons of salt annually to the lake. Deep wells were sunk to intercept the saline flow and some 60,000 tons of salt are now pumped annually and are diverted around the irrigation intake; the irrigation supply has thus been improved from 450 ppm to 250 ppm of dissolved salts.

A particularly dangerous practice in industry is the running of industrial waste into boreholes where the wastes are too concentrated to be run into rivers. Thomas and Leopold (1964) report that a factory pond at Denver, Colorado, over the years 1943 to 1961 was found to have contaminated 13 km^2 of groundwater aquifer. A major danger of contamination occurs where important bodies of fresh groundwater overlie saline waters or are in close proximity to them, as along sea coasts. Excessive pumping can then cause intrusion of the salt water into the fresh-water aquifer. This has occurred on the coast of both southern California and of Israel as a result of excessive pumping of groundwater for irrigation. Both in California and on Long Island, New York, artificial recharge is used to maintain underground levels of fresh water and thus to prevent incursion of salt water.

Estuary storage. The remaining possibility for major fresh-water storage, at the last stage before river flow is lost into the sea, is to construct a barrage across the estuary. This has several disadvantages, one of which is that the water is thereby stored at the lowest point in the catchment with maximum costs of pumping for use inland. Problems of sewage disposal and of land drainage may be serious, while both navigation and fishery rights may have to be met and provision for fish migration may be needed. Some of the estuary flats are wild-land sanctuaries for great flocks of coastal wildfowl and have strong claims to be considered as national parks. In spite of all these drawbacks, the social and financial costs of inland sites in Britain are already so high that active studies are in progress on the Wash, Morecambe Bay and the Dee (*Nature* 1972). Such proposals affect many local interests and inevitably

14

give rise to sharp controversy. The diversity of the problems has, however, engendered valuable inter-disciplinary studies in which geologists, marine biologists, ecologists, agricultural scientists and civil engineers are co-operating with the support of the Natural Environment Research Council and the Department of the Environment. This is at least reassuring evidence that a community of advanced technology is now alerted to the need to study carefully any proposals for major interference with the hydrological cycle. The opportunity to provide public outdoor leisure amenities, such as fishing, swimming, sailing or bird-watching, is now recognised as an important aspect of such schemes. In the past, such amenities have been accidental products of industry; examples are the Norfolk Broads by peat-cutting and many inland 'marinas' from gravel workings. In future they will increasingly be accepted as public objectives in water-resource development.

Desalination for domestic supplies

In the intensively developing economies of Israel and of California, both having warm climates of high evaporation rate, all the above-mentioned sources of fresh water have already been developed to a high degree and both have turned attention to mankind's ultimate water resource, the sea. Since some 97% of the water resources of the planet are denied to man by a salt content of only 3·5%, the removal of this impurity would appear to be a simple task for a technical age which sends men to the moon. The difficulties are, basically, economic; water for community use is required in very large quantities at a very low cost. The energy required to remove the salt is 0·75 kilowatt/hour for each cubic metre of sea water (about 3 kWh per 1000 gallons). The concentrated brine solution which remains is highly corrosive so that desalination plant is expensive both to build and to maintain. The techniques are already developed on a practical scale for situations in which these costs can be met, e.g. in oil-well operations in coastal deserts such as at Kuwait or where urban development has completely outrun water resources, as in Guernsey; distillation plant has been in use there for more than a decade for domestic supplies. The UN Survey of Water Desalination in Developing Countries (1964) reported a world study financed by the Ford Foundation. Nine major distillation plants with capacities of more than 1000 cubic metres per day (m^3/d) are listed. All of them used the flash distillation system originally developed in Britain. Two alternative processes had reached a plant size of 900 m^3/d, one at Kuwait using electrodialysis and the other in Israel using a vacuum freezing technique.

A recent report by the Director of the US Office of Saline Water

(Chung-Ming Wong 1972) referred to 700 de-salting plants of more than 100 m³/d capacity established throughout the world. Costs per unit volume fall rapidly with size of plant, and current costs for plants producing 5000 m³/d (about one million gallons per day) from sea water are approximately 22 cents per cubic metre (about $1 per 1000 gallons). A new de-salting plant near Tijuana in Mexico reports costs of 15 cents per cubic metre (65 cents per 1000 gallons). For brackish water the cost is below 8 cents per cubic metre (35 cents per 1000 gallons). At Siesta Key in Florida development work is now in progress on experimental plant aimed at very much higher capacities, of the order of one million m³/d. Such major plants are being considered as possible drought-insurance for major municipal systems and for processing of waste water to reduce pollution.

Two further principles are now being developed internationally, and show much promise. The first is 'reverse osmosis' or filtration under pressure, using cellulose acetate membranes with a pore-space gradient as first developed by Loeb and Sourirajan (1961) at Los Angeles. US government research developed this process and very thin films were laid down in ceramic tubes about 5 cm in diameter. This proved capable of removing organic matter as well as salts and hence offered the prospect of recovery of clean water from sewage. There is an important reduction in power demand; whereas distillation requires 540 calories per gramme and freezing needs 80 calories per gramme, reverse osmosis is accomplished at only 10 calories per gramme (Sourirajan 1970). At present the membranes are not durable enough to process sea water but pilot plants are already in operation in the UK using brackish groundwater which is available in large quantities; diluted sea water from estuaries, Trent River water and sewage effluent are also treated. Active research is in progress in the UK by the Atomic Energy Authority and the Water Pollution Research Laboratory. A second new process employing a hydrocarbon freezing agent is under development by the UK Atomic Energy Authority and Water Resources Board. A pilot plant has been successful but plans for a plant to produce 4500 m³/d have run into difficulties; a volatile hydrocarbon is mixed into cold sea water and evaporated, leaving ice and a brine slurry of dissolved solids.

Complete de-mineralisation is not often required and some brackish water is usually mixed with the pure products of distillation processes, both for economy and to improve palatability.

Solar energy for distillation. Both solar energy and brackish groundwater are abundantly available in many arid countries. Solar distillation experiments have been widely reported; those from CSIRO, Australia, appear to have reached most practical success. Brackish borehole water is lifted by a wind pump and fed into a series of black polythene troughs under pre-fabricated covers of ordinary horticultural glass panes. Evaporated water condenses on the underside of the glass and drips into narrow polythene collecting channels. A unit of 0·35 hectares is producing 11 cubic metres (2500 gallons) a day. The design is being used commercially. It is expected to be used for producing small supplies of water, up to 20 m^3/d (Morse 1967).

Atomic energy. The costs of power are such that atomic energy is the only source with the possibility of becoming cheap enough for large-scale desalination. The costs, however, are still too high for commercial development. The 'Water for Peace' conference in 1967 surveyed decisions to build 27 nuclear power plants, in the USA alone, with a capacity of 22 million kilowatts (Ramey 1967). The US government has undertaken joint planning projects with the Metropolitan Water District of California, and with the governments of both Mexico and Israel for dual-purpose combined atomic power supply and desalination schemes, but subsequent news of all three studies suggests that the costs are still too high. They are discussed by Vilentchuk (1967). Calculations for dual-purpose nuclear-energy plants producing both power and fresh water are not as yet based on any large-scale experience. The service charges for capital are a critical factor. Preliminary exercises for the America-Israel scheme for a large (250,000 m^3/d) plant suggest that capital charges at $4\frac{1}{2}\%$ would constitute half of the cost of the water; thus even the economies of very large scale can be offset by small changes in the service terms for capital costs.

The geographical limitations of coastal supplies are also important and the production of large quantities of water at a few coastal sites can do little to meet the everyday needs of human populations scattered over millions of square miles of large continents.

Desalination for irrigation

The geographical proximity of the oceans to irrigable deserts provides the theoretically optimum agricultural sites for such nuclear-energy desalination plants. Unlimited water and land are thus available along some 32,000 km of coastal deserts where sunshine and temperature are not limiting factors for crop growth (Meig 1966). The attractive picture of large-scale dual-purpose nuclear coastal installations supplying de-

17

salted water direct to very large-scale and well-equipped agricultural undertakings, may at first sound like science fiction, but Hammond (1967) has made a case for the economic feasibility of such schemes under favourable circumstances, even allowing for 7% fixed charges on capital. Farm operations would need to be at high efficiency with skilled staff making the maximum use of agricultural science and with highly efficient plant maintenance. Although Hammond's estimates are more optimistic than those reported for the America–Israel project, there is little doubt that such schemes are, scientifically and technically, worth investigation on an international scale. Weinberg (1969) describes encouraging multi-disciplinary studies at the Oak Ridge National Atomic Energy Laboratories from which it is suggested that very high irrigation efficiencies could offset higher costs of water supplies in such integrated and scientifically controlled developments. The social problems of associating desert nomads or primitive cultivators with such exacting advanced agricultural techniques would need to be undertaken in schemes which are financed and costed separately.

The present prices of atomic power output are based on the supply of naturally occurring ores of fissionable uranium-235. This is only about 1% of the uranium as mined and the supplies are limited. In the quantities needed for water production they would have a life of only a few decades. Breeder-reactors capable of rendering the whole of the uranium or thorium active, and thus using the other 99% of the ore content, have begun production but are as yet few in number (US National Academy of Sciences 1969).

At a recent conference on desalination, held in Madrid by the International Atomic Energy Agency, the conclusion reached was that atomic power at costs low enough for large-scale water supply would not be attained for at least a decade.

Oil-fuelled processes. Desalination of sea water is a popular cause with those who oppose the 'drowning' of more valleys for water storage to supply growing cities, but the stage is not yet in sight at which the sea offers a practical alternative to conventional sources of water supply except in the most extreme circumstances.

Although the atomic-powered desert irrigation plant is a distant prospect, a more modest scheme has already reached pilot-scale development. The Environmental Research Laboratory, a unit of the Institute of Atmospheric Physics at the University of Arizona, has set up a small working unit on the north-west coast of Mexico (Hodges and Hodge 1971). Using, as energy input, the waste heat from the diesel motors of the electricity-generating plant for the local community, together with a modest power supply for pumps and fans, heat exchange between counter-currents of air and water is used ingeniously to extract about one litre of fresh water from every ten litres of sea water. The brine is discarded back into the ocean. At present the pilot plant distils 10,000 litrès (2400 gallons) a day. Operational costs will depend on how successfully the problems of the corrosive action of hot sea water on metals have been overcome.

Such water is too precious to use in open-air irrigation, but greenhouses, covered with polyethylene and maintained by inflation, have been built beside the desalination plant. All this may read like science fiction, but 85 varieties of vegetables, including 18 different species, have already been grown with success (Jensen and Teran 1971). The authors report that the Shaikh of Abu Dhabi, in the Arabian Peninsula, has given the university at Tucson a grant to develop a power–water–horticultural station at Sadiyat in his country.

A recent progress report (Frenchman 1971) describes rapid progress in a full-scale experiment, with additional financial support from the Rockefeller Fund and other sources. Plate 3 illustrates both the plant at

Plate 3. A pilot-scale irrigation plant using desalted sea water.
(Photograph by Capt. Hewson, Gulf Aviation.)

19

Sadiyat and the genuinely desert nature of the terrain. Yields higher than average for glasshouse crops are already reported and 170 different vegetables are under test.

Basis for comparison of costs. In seeking some perspective for the costs of future public water supplies for the more industrially advanced communities it is necessary to remember that the most readily available water sources and the most geographically favourable storage sites have, in general, already been used. They are in part responsible for the very low water costs which are a feature of modern industrial communities.

It is most important, however, in considering the costs of any new proposals for water supply, that these should be compared, not with the supply of water from systems established at lower costs in the past, but with supply from new developments, by alternative conventional methods, in the future.

Rain-making

Means to increase precipitation over watersheds would be of the most direct importance to water-resource development. Watching the towering cloud masses in which thousands of tons of water droplets sail serenely over parched countrysides, man's urge to stimulate rainfall has been acute and widespread. In primitive societies priests or rain-makers propitiate their gods, while in modern technical times farmers call on governments to apply science to the problem. Some progress has been made by direct attacks on the clouds with rockets or with aircraft and after several decades of experiments it is now clear that favourable responses are limited to very special circumstances. Condensation of water vapour, into drops heavy enough to fall as rain, begins around very small nuclei, such as airborne particles of dust or fragments of ice. Clean air, without such nuclei, may become supersaturated or overladen with water vapour. In such conditions man can intervene effectively to supply suitable nuclei around which condensation can begin. These techniques are known as 'cloud-seeding'. They are useful only in clouds where the water content is high enough and the temperature is low enough to be close to the conditions for precipitation. 'Seeding' with particles of common salt, silver iodide, or 'dry ice' (carbon dioxide gas frozen into solid particles) can then cause heavy precipitation. Unfortunately, the conditions suitable for precipitation do not often occur in dry weather when extra rain is most critically needed. Nine twin-engined aircraft were flown for three years in cloud-seeding experiments over the Snowy Mountains Hydro-electric Scheme in Eastern Australia. Rainfall increases of 30% were observed in this high-rainfall area but none at all were achieved in similar experiments over drier country in South Australia (Bowen 1966; Smith 1971). 'Seeding', either from aircraft or by the firing of the rockets from the ground, has therefore achieved only limited practical success. Indeed the firing of rockets often does more to relieve the feelings of drought-stricken farmers than to give relief to the crops.

Artificial stimulation of rainfall therefore offers no solution to the problems of semi-arid climates or of major droughts in the farming areas of the world, although continued experimentation may locate more of the special areas in which it is useful.

Greater progress has been made in techniques of 'cloud-seeding' for the prevention of hailstorms. Here the problem is to detect critical cloud formations which are approaching areas in which valuable crops are at hazard. Premature precipitation of the water load is then brought about by 'seeding' the clouds before they reach the crop area and before conditions for the generation of hail are reached. Organisations of growers maintain active field defences for their vineyards in Italy, for tea plantations in Kenya, for tobacco plantations in Rhodesia and for other valuable crops in the USA, Australia, South Africa and Israel. Thus although the 'cloud-seeding' technique will not produce rain in a drought it can provide some practical protection from destructive hailstorms.

Summary of resources

Most of the world's water is too salt for use. The fraction of the fresh water available to man is only about one two-hundredth of the world's total store. We can do little or nothing to control the circulation of water in the atmosphere and man's influence is exerted only on the reception and distribution of the precipitation which reaches the land surface. By our management of the land we can often ensure its absorption as clean water into the soil and we can control, to some extent, its flow towards the oceans, its storage and its distribution. Above all we have the task of ensuring that the water remains free from pollution so that it is serviceable for the multiple uses of our human communities and for our many neighbours of other species.

There are prospects of winning fresh water from the sea by nuclear power at a few critical places where the great costs can be justified, but little prospect that this source may ever become a main supply on a world scale.

2

Development of a watershed discipline

The historical choice of discipline or disaster

In man's early history only irrigation systems appear to have been able to shape land and water development on a logical basis. In some cases they have imposed an overall discipline with great powers of survival. Writing of 4000 years of irrigation history in Mesopotamia from 2400 BC to 1600 AD, Jacobsen (1958) concluded that 'a particularly close relationship exists between the flourishing of irrigation agriculture and the existence of a stable and vigorous central government'. Large-scale irrigation farming continued for thirty centuries between the Euphrates and the Tigris rivers in Mesopotamia and there is evidence of many centuries of prosperous irrigation in Ceylon. Both ended in destruction through inability to maintain the essential discipline.

The Beled Dam on the River Tigris upstream from Baghdad diverted flood waters into a large impervious flood basin. The Nahrwan Canal carried irrigation water to a network of channels which drained into the Euphrates (Willcocks 1911). Failure to maintain the heavy task of clearing the channels of soil from torrent flows, probably accentuated by misuse of the headwater catchments,* appears to have led to the river breaching its banks and changing course. Ruined cities, in a waste of drifting sand, are all that remain (Bennett 1939). Erosion of the headwaters continues as populations of subsistence agriculturalists and graziers increase. In the Tigris flood of March 1953 the river was estimated to have carried 14 million tons of soil and rock debris past Baghdad in one day (West 1958).

The earliest irrigation works in Ceylon are believed to have begun in 2100 BC. By 1100 AD there were some 15,000 storage tanks behind earthen embankments which can still be identified. The largest reservoir

---

* The terms *river basin, drainage basin, catchment area, watershed*, are used to describe the land surface from which water flows to a given point on a watercourse. The limits of such areas are described as *boundaries, perimeters* or *drainage-divides*.

(the 'Giant's Tank') covered an area of 2400 ha (6000 acres) with over 8 km of stone-faced earthworks. Warfare and malaria subsequently destroyed the irrigation communities and most of the earthworks were breached and overgrown by forest (Brohier 1934). In contrast, maintenance of irrigation discipline over many centuries in Peru has continued the use of the elaborate stone-walled pre-Inca terraces to the present day.

Organisation for mutual protection on a watershed area

Floods. Floods are the most dramatic symptoms of land misuse, but not all floods are due to man's mismanagement. Monsoon climates concentrate heavy rainfall into a few months, so that seasonal torrents have poured from the Himalayas on a geological time-scale to carve their spectacular canyons through the lower country. Even in temperate climates occasional random movements of converging air masses cause unpredictable excesses of rainfall, overloading a countryside with water, and producing floods in landscapes of gentle topography, stable soils and good agriculture.

The village of Lynmouth in North Devon was largely destroyed in 1953 by a flood in which twenty-eight people were drowned and seventeen bridges were swept away. It was produced by continuous heavy rainfall, none of which was exceptional, but the rain accumulated continuously in a quite unusual manner so that the catchment of only 39 square miles produced a flood which the Thames rarely attains from a catchment one hundred times larger. Fortunately, in such country the probability of recurrence is low.

Although Great Britain experiences only one tornado a year as compared with 143 a year in the USA, the need for flood prevention increases as the population grows and the network of urban areas becomes more dense (*Nature* 1953). On the recommendations of the Institution of Civil Engineers, a special flood study project has been set up by the Natural Environment Research Council as part of the programme of the Institute of Hydrology.

Soil conservation. Where land is severely over-grazed or over-cultivated no such exceptional rainfall is needed to visit the sins of residents in the upper catchment on those lower in the valley, as the early farmers of the USA learned to their cost. The organised community discipline of the Conservation Area, by which the vote of a majority of landholders within a catchment basin can secure the application of in-

23

tensive conservation laws throughout the area, was established in the USA by the Soil Conservation Act of 1935; it has become the basis of agricultural land-use policy throughout the country (Bennett 1939; USDA 1970).

The basic tenets of the soil conservation discipline are that the soil surface should be maintained in a receptive condition for the infiltration of rainfall and that surplus water should be led off along gentle gradients without reaching erosive velocities. Where a steep fall is necessary, concrete channels or other revetment should be provided.

The surplus water should be guided along prepared drainage routes which should be maintained with an erosion-resistant surface, such as permanent grass, and be kept free of obstructions. The drainage routes ignore the boundaries of individual properties and serve the valley as a whole, so that the co-operation of all of the farmers is essential to success. Such local organisation depends for success on a strong Agricultural Advisory or Extension Service and offers, of course, an excellent channel of contact between government and farmers. The rules usually insist on cultivation on the contour, the construction of cut-off drains and of grassed waterways, with terraces where necessary for the steeper arable slopes and land drainage of the flatter areas. Some element of subsidy for such structures, including farm storage reservoirs, is provided in many developed countries, even where agriculture is at a high level of prosperity. For developing countries subsidies for such works are quite essential to progress. In Southern Africa this system has proved successful in advanced freehold farming areas from the Equator to the Cape and is making slow but definite progress in the communally held tribal areas.

Contour farming is little practised in Northern Europe and the UK, because rainfall exceeds evaporation and is well distributed throughout the year. Drainage, particularly for heavy soil, becomes the dominant factor so that ploughing down the slope is traditional, and safe. Most of the world's agricultural population lives, however, under soil conditions in which ploughing down the slope leads to agricultural disaster. Where evaporation exceeds rainfall, there are characteristically high rainfall intensities and the frail kaolinitic clays of the tropics and sub-tropics erode rapidly if misused.

Economic progress. The main sources of support for a land-use discipline are, thus, firstly irrigation and secondly protection from floods and erosion. The third, and by far the greatest force in current operation, is the pressure of economic competition as agriculture reaches the level

24

of the cash economy. Erosion is rarely tolerated in soil to which expensive seeds and fertiliser have been committed, while the embarrassments of soil-laden water and the failures of perennial streams are not passively accepted by a community aware that the means for remedy are available. Progressively, as the standard of living rises beyond the subsistence level, more active community interest is taken in organising prevention of soil and water mismanagement.

Protection of streamsource areas

Projecting features of a landscape, whether mountains, hill ranges or the escarpments of plateaux, serve to force approaching airstreams up to cooler altitudes, at which their water vapour condenses into rain: they are therefore commonly source areas for streamflow. Trees, as plants best able to make use of abundant water to overgrow and dominate rival species of lesser stature, are commonly the ecological climax of vegetation of such high ground. Forest Reservations are therefore a usual form of protection for these important streamsource areas. This is readily understood by technically developed communities, and in many developing countries the governments have, to their great credit, taken sound advice and established Watershed Protection Reservations. These are difficult to explain to peasant farmers on a precarious subsistence level of agriculture, since they see the Forest Reservation boundary only as a denial of grazing opportunity for their livestock, or a with-holding of fertile forest soil from their crops. Their marginal economic situation does not incline them to show interest in the fate of others living downstream. They do not yet know or care about the importance of regulated flows of clean water for the prosperity of the country and its future generations. *This is a case where community discipline must hold the boundary while community education catches up. Persuasion is always the best approach, and offers the only permanent solution, but no country can afford the destruction of irreplaceable natural resources for lack of protection during the long process of public education.*

Just how long this process can take, and the extraordinary resistance of the human race to the lessons of our environment, are brought home sharply by the following quotation from Plato (*Criteas*, about 400 BC).

'There are mountains in Attica which can now keep nothing more than bees, but which were clothed not so very long ago with fine trees, producing timber suitable for roofing the largest buildings; the roofs hewn from this timber are still in existence. There were also many lofty cultivated trees, while the country produced bountiful pastures for cattle. The annual supply of rainfall was not then lost, as it is at present, through being allowed to flow over a denuded surface to the sea. It was received by the country in all its abundance, stored in impervious potter's earth, and so was able to discharge the drainage of the hills into the hollows in the form of springs or rivers with an abundant volume and a wide distribution. The shrines that survive to the present day on the sites of extinct water supplies are evidence for the correctness of my present hypothesis.'

It is indeed humbling to realise that so clear an analysis was possible by observation and deduction four centuries before the Christian era. It is also alarming that more than 2000 years later at least half of our human race lives in rapidly growing communities which make no effective provision for the protection of watersheds.

The desolation of which Plato justly complained persists in the overgrazed mountains of Greece and of Turkey to the present day, although remedial action, by the replanting of forests, is slowly progressing. Plate 4 illustrates the two extremes. The penalties are paid over the centuries, as lowlands are flooded by torrents from the bare mountains and as seaports are left far inland by the advancing deposits of soil and rubble washed down from the hills.

Outside the protective boundaries of watershed reservations, man has not yet evolved a universal set of laws to govern even such elementary matters as the abstraction of water from a river or the discharge of wastes into it. British and Spanish legal traditions, for instance, differ sharply in this respect. The differences are most apparent in mid-western USA where Spanish water law, under which the west coast was originally settled, meets water laws in the British tradition which have spread from the eastern states.

Critical areas of erosion hazard. The need for watershed discipline to be consistent throughout the whole area of large watersheds is often more apparent to the engineers than to the civic authorities. The latter tend to be reluctant to spend substantial funds, and to fight political battles, for control of land use in remote areas difficult of access. The extent to which small areas of extreme erosion hazard can damage the hydrological conditions of a watershed is well illustrated by the current situation on the Parana River. The following details are quoted from the UNESCO reports of the Mid-Decade Conference of the International Hydrological Decade (Pereira 1969).

In the steep and semi-arid slopes of the high Andes, soil erosion from the headwaters of the Bermejo River tributary, which constitutes only some 4% of the watershed of the Parana, causes serious economic damage to the major community of Buenos Aires.

Plate 4a and b. Overgrazing and restoration.

Two contrasting views taken from the same spot. Plate 4a is typical of the overgrazed condition of large areas of mountainside in Turkey. Plate 4b shows active restoration by reforestation. (Photographs by William Allan.)

The Parana is a truly international river, rising in the Matto Grosso of Brazil and flowing southwards more than 2000 km to Buenos Aires and the River Plate estuary. On the way it collects tributaries from Bolivia, Paraguay and northern Argentina, draining a total area of over

1,000,000 km². The overall gradient of the major area of the watershed is very slight and flows from a high proportion of the area filter slowly through marshes from which they emerge free from sediment. After 800 km of its journey through Brazil, Bolivia and Paraguay, the main channel, here called the Paraguay River, has an average flow of 4000 m³ per second of clear water and represents for Argentina a major national resource. This fine river is, however, joined about 10 km north of Corrientos by the muddy flow of the Bermejo River, draining the overgrazed mountain provinces of Salta and Jujuy at the extreme northern end of Argentina. Sediment transport studies show that the Bermejo River contributes some 80,000,000 metric tons of suspended sediment each year. The material is a clay of extremely fine particle size (less than 15 microns) and is mineralogically identifiable. These fine particles remain in suspension for the next 1200 km until flocculated and precipitated by sea water.

The deposition of this sediment is a serious matter for the port of Buenos Aires and for the major river transport system of northern Argentina. The Parana River is navigable for 400 km up to the ports of Rosario and Santa Fe for ships of 15,000 tons. Smaller ships of 5000 tons can navigate seasonally a further 600 km to Asuncion. The estuary of the River Plate is wide and shallow so that dredging is needed to maintain a channel for 100 km seawards of Buenos Aires. The annual deposition of 100,000,000 tons of silt requires constant dredging which at present costs about US$10,000,000 a year. Mineralogical studies by the Authority for Water and Power show that some 80% of the 100,000,000 tons a year of sediment comes from the Bermejo River headwaters.

This high steep country lies between 300 and 2000 m in altitude. At its lower edge the river gauge at Zanja del Tigre (altitude 296 m) records flow from a catchment of only 25,000 km², yet the total annual flow of suspended sediment is 65,000,000 tons. This is equivalent to a loss of over 2500 tons per km². The catchment area of the San Francisco branch of the Bermejo above the river gauge at Urundel (altitude 367 m) is about 26,000 km² and contributes about 15,000,000 tons per annum, thus losing 560 tons/km²/year. This small mountain area, in total only 4% of the vast Parana Basin, thus contributes 80 out of the 100 million tons per annum of deposit. The land use is described as 'mountain pasture with many goats in a low rainfall zone' and a report published by the Ministry of Agriculture of Argentina in 1957 shows the Salta province as an area of severe soil erosion.

It may be significant that it was in the Bermejo Valley more than four centuries ago that the cattle industry of Argentina was founded by the arrival of a small consignment of eight Andalusian beasts in 1556. By 1588 there is a record of the export by a rancher named Hernandarias of the first three thousand head of cattle and fifteen hundred horses by swimming them across the Parana River to Corrientes. The grazing effects of four centuries in low and strongly seasonal rainfall are not yet controlled by a soil-conservation service.

Severe soil erosion is characteristically an accelerating process. There is a well-equipped commission (CLIAP) studying the trends. Soil is deposited to build new land surface in the delta at a rate of 3 km^2 each year. A series of surveys shows clearly that from 1873 to 1879 the delta was advancing at the rate of 46 m per annum, while the rate from 1900 to 1964 was 84 m per annum. There is no sign of an increase in river flow to account for this and indeed there are some indications of a decreasing trend of flow.

Apart from the serious implications for a major sea port and a major system of inland water transport this unwelcome contribution by the Bermejo River is embarrassing to the city water-supply authorities. The fine suspended clay is difficult to deal with in water-treatment systems and the city of Buenos Aires spends some US$2,000,000 per annum on chemical flocculation treatment of water for the 6,000,000 inhabitants.

The Comision Nacional del Rio Bermejo was set up to remedy this situation and a civil engineer's report (Cotta 1963) ascribes the main volume of erosion to bank-cutting by torrent flows. Construction of embankments for flood diversion and protection of the eroding areas is the essential first stage in the restoration of watershed control, but in the long term the correction of land use by the protection of steep slopes from grazing and trampling will be essential to success.

Multiple land-use planning to include recreation opportunities

Urban concentrations. An increasing drift of people into large urban concentrations is a feature common to countries at all stages of economic development. The water requirements of these congested areas grow alarming, and strain the geographical resources of the countryside. From the UN Headquarters, the Department for Economic and Social Affairs gives the startling forecast that in a single current lifetime, from 1920 to 2000, the world's *urban* population is expected to increase *eightfold*, from 360 million to some 3000 million. Half of the total world population will live in cities or in larger metropolitan areas, some of which will have merged into 'urban regions'. In the most developed countries four-fifths of the population will be urban residents.

Public access to protected catchments. Around these centres of concentration, for which the term 'conurbations' is gaining currency, intensive agricultural development already restricts the opportunities for recreation in the countryside. Access to unspoilt areas of country of scenic beauty, or at least of spaciousness and peace, is now a valued part of a high standard of living. As populations grow, these opportunities for refreshment of spirit may well become psychological necessities for the majority, as they already are for a minority, of the dwellers in a concrete environment. The scenic values of water-source areas of high ground therefore permit the dual roles of watershed protection and of recreation to be combined. With modern methods of water purification public access may safely be allowed to watersheds and only the areas immediately surrounding treatment works and subsequent distribution reservoirs need exclusive protection. The dual protection of water resources and provision of recreation takes many forms such as National Parks, Game Parks, Forest Reservations, Wilderness Areas, Game Sanctuaries, Hunting Reservations and special protection of unique habitats, as for waterfowl or for rare species of birds, animals or plants. Management techniques are being evolved to secure multiple use of such land resources, in order to produce timber or meat as well as water, without destroying recreation values. Studies in the USA by the Forest Service (Hewlett and Douglass 1968) and in Africa by the United Nations (Swift *et al.* 1963) illustrate the problem in detail.

Management of wildlife populations. The essential control of wildlife populations, to prevent them from over-running and destroying their own environment, may make harvesting of animal products incidental to good management. Rapid deterioration of vegetation, soils and water supplies can result from overstocking with wildlife as with domestic stock (Plate 5). Striking evidence for this is available from wildlife protection areas in the USA (Croft and Ellison 1960; Plummer, Christensen and Monsen 1968) and in Rhodesia (Savory 1965). In East and Central Africa, for instance, where the rich variety of wild animals forms a major tourist attraction and is the basis of economically important tourist industries, management of wild-lands has become recognised as needing a highly specialised skill, supported by considerable capital equipment and staff. Populations of major browsers and grazers, such as elephant and buffalo, must be controlled, and the delicate balance of grass and bush in the savannah country must be maintained by the distribution of water points and control of fire. Highly organised hunting and harvesting of meat, hides, and byproducts from elephant, hippo-

30

Plate 5. Overgrazing by both cattle and wild species.
African buffalo on overgrazed rangeland. (Savory 1965.)

potamus, buffalo and many varieties of lesser game have to be under-taken in remote parts of the major parks, inaccessible to the public, by staff equipped with aircraft, cross-country transport, cold stores and weapons which strike with tranquilising drugs. Valuable and sustained sources of high-class protein for local consumption are thus provided in Kenya, Uganda, Zambia, Rhodesia and South Africa (Talbot *et al.* 1965; Ledger 1965), but the areas are usually in dry warm country rather than on high-rainfall watersheds (an illustrated discussion is given by Pereira 1964). In the USA the preferred method of controlling the large populations of wild deer in watershed protection forests is to open the areas to licensed public hunting.

Management for preservation of scenery. A close and inexpensive integration of land uses is to be found in the National Trust Lands of Great Britain. Watershed areas of hill grazing and mature parkland under stable farming management, including buildings of historical and architectural interest, are opened to public access without interruption of their agricultural use, but are preserved from further building or development which would destroy their scenic and recreational value. Cherished parklands and stately mansions, whose owners are no longer able to maintain them against rising costs and levelling taxation, are given to or shared with the National Trust for preservation as public assets. The management of these steadily increasing areas of publicly

31

protected parklands and farmlands is a matter of some importance for water-resource conservation as well as for public recreation. Intensive modern agricultural methods are usually less aesthetically attractive and can afford far less tolerance of public access than the traditional farmlands. If left unmanaged, the fields revert rapidly to thicket and bramble, with reduced access and scenic value. On poor heathland this is acceptable, but on better land the continuation of grazing provides a favourable watershed cover, as discussed in later chapters, for the acceptance and collection of clean water. As arable cropping becomes increasingly concentrated under intensive capital development on the better soils, it will become possible to increase the areas maintained for public access as grazed parkland and thus to improve both water supplies and amenities.

Management for maximum water output

Steadily dominating all aspects of multiple-use management of water-source areas in the more developed countries are the constantly rising demands for more water. Both in large countries such as the USA and in small and already heavily populated countries such as the UK, Belgium and the Netherlands, sums of substantial importance in the national economy are committed to development of water supplies for the major cities, as these out-grow the geographical resources on which they were originally founded. Large volumes of water of good quality and under hydraulic control are therefore a saleable product, and if held high in the river systems, their potential energy, recoverable in part as hydro-electric power, is also saleable.

A discipline increasingly imposed as populations increase and the standards of living rise, is the combined use of land not only for agriculture, forestry or urban development but also for the safe reception and delivery of an unpolluted harvest of fresh water. Essential criteria must therefore be the ability to produce an orderly flow rather than a raging torrent and to minimise losses by evaporation and pollution. The extent to which these criteria are met by forestry, by agriculture and by urban development is discussed later.

The Tennessee Valley Authority is a highly successful example of a pioneer social and engineering venture for the restoration of a complete watershed, of 41,000 square miles, from conditions of poverty and erosion to those of present prosperity. Nine great multi-purpose reservoirs, backed by many smaller storage dams, provide water for power generation, for irrigation and for the maintenance of a 650-mile navi-

32

gation canal as simultaneous objectives which have benefited some seven million people. Half of the area is protected by permanent forest on steep land, while critical hill slopes in farming areas are also under forest reserve. Some of the forested areas are developed for tourism and recreation, while the farming areas are organised on a watershed basis into seventeen Development Associations.

The most striking and imaginative demonstration of the harnessing of natural watersheds is to be found in the Australian Alps, where the Snowy River scheme produces both electric power and irrigation water. Surface runoff is collected among rugged mountains into seven large reservoirs, linked by tunnels serving three power stations, so that water which formerly flowed eastwards into the sea, now traverses the mountain range, yielding power on the way, and flows into the major irrigation schemes of the semi-arid Murray River valley (Chapter 9). Results of land-use research from both these schemes will be quoted in later chapters.

As demands for water and for power continue to grow an increasing land area will be pre-empted for water-resource development. The high economic and social cost of land for reservoirs has already been mentioned.

Increasing investment at the study and design stage is therefore justified, and fortunately a new technology has provided help. Simulation techniques with the digital computer have transformed the study of water-resource development in the past five years. Instead of designing single reservoirs it is now possible to undertake the designing of river-regulation systems in which supply and demand are matched by means of storage in reservoirs and sometimes by pumping of water between reservoirs. Scanty and fragmentary data of river flow and rainfall over the watershed are now assembled in mathematical models which make the best use of all the information available. Exploration of the possible effects of rainfall occurrences, sequences and distributions can then be extended by tests with artificial populations of rainfall data generated from random numbers. On the model of the normal river flow thus constructed it is possible to superimpose the estimated future demands of urban water supplies drawn from several points in the system. Reservoirs can thus be planned as parts of a water-resource network rather than as single projects.

In steep countryside, subject to heavy rainfall, spare capacity behind additional dams is needed for water control alone. The dams must be maintained partly empty in order to hold and delay the peak flows in separate tributaries and thus to prevent their coincidence in the main

33

river. The US Congress passed flood control measures in 1936, 1944 and 1954 which have given to the Soil Conservation Service the responsibility for a nation-wide programme of such constructions. In Texas alone 1300 structures, delaying runoff from 5200 square miles, have been built since 1950 and a total of 3500 are planned. Water losses by evaporation and seepage from these reservoirs may be serious, amounting to some 10% of the yield in watersheds producing an annual runoff of 250 mm, but rising to 70% or more in drier country with annual runoff of only 25 mm (Sauer and Masch 1968).

An example of the hydrological effects of such flood-retarding structures is given by Kautz (1955). In the restoration of an eroded watershed of 260 km^2 area, 24 such flood-delaying dams were built, the water levels being drawn down by controlled outlet pipes. Each such structure checked the flow from about four square miles and was able to hold back the maximum runoff to be expected in 25 years. Grassed spillways bypassed major flows up to the maximum to be expected every 100 years, and the volume of storage allowed for some 50 years of deposition of silt before the dams became ineffective. Detailed studies of the watershed before and after the construction of the 24 dams showed an 85% reduction in total flood damage after severe storms.

The flood-control effects of a very thorough conservation treatment of the steeply divided watershed of Sugar Creek in Oklahoma were reported by Hartman et al. (1967). After seven years of autographic records of streamflow and of rainfall, minor gulleys were stopped and 25 floodwater-retarding storage structures were built in the upper third of the 620 km^2 (240 sq. mi.) valley. Their total capacity was $18 \cdot 5 \times 10^6$ m^3 (15,000 acre-ft).

This conservation development halved the floods from a subsequent severe 200-mm storm.

Conservation v. collection. Even the best traditional land uses may be inadequate in climates where evaporation exceeds rainfall, i.e. under the conditions in which the majority of humanity lives. The greater the success of the soil conservation specialists against erosion, by helping farmers to prevent runoff from their fields, the lower fall the levels of water in the reservoirs downstream. This has been a sharp problem in South Africa, where improved soil conservation in the catchments supplying Johannesburg has greatly reduced both the high sediment load ($0 \cdot 8\%$ of total flow) and also the catch of the reservoirs serving the city. The developments of the Vaal and Orange Rivers, together spanning a continent, are typical of the vast engineering works essential

34

for raising the living standards of growing populations in climates of water deficit. Both soil conservation and increased supply of water are needed (Olivier 1962, Midgley 1963).

In the USA this problem has been severe enough to attract public funds to studies of the ultimate solution, the allocation of land solely to the reception of rainfall. The US Water Conservation Laboratory has been working for several years on methods of sealing the soil surface with silicone resins or with sprayed emulsions of asphalt (Myers 1967). Plastic sheets protected by gravel, and also aluminium sheeting bonded to a sprayed asphalt surface, are under test. In Hawaii catchments of up to 7 ha have been covered by artificial rubber sheeting for the collection of water. Myers (1967) estimates practical costs at less than 20 US cents per cubic metre of water. Since high rainfall and high ground often coincide, well selected sites permit gravity conveyance to storage reservoirs although there may well be aesthetic objections to the effects on the scenery. More study will undoubtedly be given to water harvesting from sealed surfaces in semi-arid country. Although the materials are new the device is very old indeed. Professor Evanari and his colleagues (1968) have successfully reconstructed some of the water-harvesting plots and runoff channels left by the Nabathean peoples in the Negev Desert some 3000 years ago.

In a formidably dry and hot wilderness, with a surface of 83% of bare limestone rock, 14% of barren flats of loess and 3% of steep wadis, erratic winter rainstorms over five months average 90 mm. By channelling runoff into small pockets of deep soil, fruit trees have been sustained to crop annually and perennial range grasses have given high yields (25–35 t/ha) on the small irrigated areas (Shanan *et al.* 1970).

Water loss by evaporation from reservoirs. Part of the discipline imposed by water shortage must be that of increasing the efficiency of storage. Shallow pools behind small bunds in hot summers dry up rapidly and can cause serious water losses. Langbein (1962) quotes the case of the Cheyenne River basin, where half of the total flow is trapped in ponds for watering of stock, with evaporation losses amounting to one-fifth of the river flow. Evaporation losses from reservoirs are a continuous challenge to water engineers and have received substantial research attention for two decades, but with only limited success. Great hopes were aroused by experiments with films of a heavy alcohol, hexadecanol, a waxy substance which spreads out on water to form films only one molecule thick. On small tanks when the water surface is calm the film persists and evaporation is halved. On larger areas the

35

film is broken by wave action so that for small reservoirs of up to about 50 ha, the evaporation is reduced by only 30%. A very large-scale study was undertaken in the USA in 1951–2 and again in 1956–8. On both occasions attempts were made to maintain a film over the 1000-ha surface of Lake Hefner: three major government departments joined with the Oklahoma State and Civic Authorities (US Geological Survey 1952; Lake Hefner, Report by the Collaborators 1958). The hexadecanol film did not prove to be strong enough and only a 10% reduction in evaporation was achieved (Harbeck and Koberg 1959). Development of this technique now appears to depend on whether the chemists can devise a stronger film. For small reservoirs there has been an interesting development of a new technique in the USA, Australia and South Africa. Floating hexagonal panels about one square metre in area are cast from expanded polystyrene. Tests have been successful but the costs are not yet low enough for routine use.

Community organisation by watershed boundaries. Developments of this nature will continue to improve the techniques of the water engineer but the organisation of the land use and the choice of priorities in the care of water resources concern the community as a whole. Discipline involves an orderly structure of society: for water affairs this dictates some reference to main catchment-area boundaries in ordering the hierarchy of local and regional responsibilities. In countries with long civic histories and little previous need to be concerned with water affairs, this has required some drastic reorganisation. In Britain it has proved to be a very slow process. The principle of a unified management of catchments achieved a substantial degree of administrative practice only in 1965 with the setting up of the River Authorities and the Water Resources Board; this Board is now responsible for national planning of the development of water resources, although it has to report to two separate Ministries and to the Secretary of State for Wales. Further reorganisation is in progress to concentrate authority in ten regional boards.

In newly-developing countries local communications can sometimes prove to be a difficulty in the organising of agricultural communities on to a catchment area basis, since the roads run mainly along the drainage divides and bridges may be far apart. Rivers are often boundaries of tribal or clan areas, but even where they are also linguistic or national boundaries the logic of watershed discipline is inescapable if both sides are to develop the water resources of their common valley.

The UN technical agencies, backed by the UN Special Fund, can catalyse this co-operation across national boundaries within a river basin by offering technical help in regional schemes on a watershed basis. Progress is being made in such schemes for the development of the resources of the Chad Basin in West Africa and of the Upper Paraguay Basin in South America: both schemes involve four neighbouring countries.

It is perhaps fortunate that the vast costs of modern water-resource development are forcing the most technically advanced communities to look more critically at their use of water. By tradition it has been a cheap commodity, which has been described as 'collected, stored, purified and delivered at a few cents a ton'. The relentless increase in the water use by industry, noted in Chapter 1, must eventually outrun all possibility of further supply. Recirculation, treatment and re-use are often possible but are expensive. Clearly the corollary of watershed discipline by the rural community must be the more disciplined use of water by their urban and industrial fellow citizens.

3

The achievement of hydrological information

The first stage in research and development of water resources

The variable input. Photography from space vehicles, presented in the public Press, has shown a vivid series of pictures of cloud patterns covering vast areas of the world's surface. From the swirling shapes of cyclonic storms, the dense concentrations of visible water vapour over some areas and its apparent absence over others, the general public can now readily understand that the rain and snow falling from this pattern must be highly variable in time and space. Its measurement cannot therefore be simple or cheap. It is clear that a pattern of rain-gauges rather than a single gauge must be needed over any area of practical importance, such as a city, a farm or a watershed.

In public debate it should now be easy to present a convincing case for adequate funds for the measurement of rain or snow and of its effects on the flood and drought of our streams, reservoirs and underground waters. These funds have been inadequately provided in the past in many technically advanced communities. In Britain, for instance, only five rivers, including the Thames, have detailed charts giving a fifty-year run of flow records. Seventy more have twenty-five-year records, although over 400 now have six years of record. This recent improvement has arisen from growing concern about the future trends of water supply and of rising demand in Britain.

The variability from year to year is confirmed by everyman's experience, but the corollary that measurements can be useful only over a run of years, so that future development will need measurements which must begin now, requires a degree of foresight which has lagged sadly behind man's other technical progress.

The dividends of foresight. In developing countries, where much of the infrastructure is still being designed, there is a very strong economic argument in favour of the earliest possible establishment of continuing hydrological measurements. The engineers who will design the storage dams, irrigation systems, road bridges, and even road culverts and storm-water drains, must make sure that their structures are successful. In the absence of information as to rainfall and streamflow, they must build with costly increases of safety margins against unknown perils. These increases are likely to cost more than would be needed for a decade of simple routine data collection. There are also good reasons for the early mounting of groundwater reconnaissances, lest costly dams be built and pipelines laid in ignorance of cleaner and cheaper underground water nearby.

The world's shrinking reserves of empty, uninhabited lands now present the major problem. They are often critically important sources of clean water; very large areas, as in Western Australia and in much of Africa, are scheduled for agricultural development to supply crops for the world's accelerating increase in human population. Decisions on land use, the protection of water-source areas, the allocation of resources to agricultural development and to the costly infrastructure to support it, all depend critically on estimates of the hydrological regimes ruling in the empty lands. The lack of roads and habitation has hitherto been decisive, and for most semi-arid areas there are no direct measurements whatever. As Walter Langbein (1962) so succinctly states, 'Water facts are where water is.' This picture should already be changing fast, for we now have the means at our disposal. Research and development of automatic weather stations and remote recording devices have resulted in practical designs, now under field test and initial routine use in the USA, in the USSR, in Australia, in the UK, and in many other countries. Radio communications and transport by the four-wheel-drive cross-country vehicle, the light aeroplane and the helicopter have made installations and maintenance possible in remote and roadless areas. Rapid mapping by stereo-plotting from aerial photographs not only reveals topographic features, but also much information about vegetation. Deductions about geological structures and soil types, surface flows and the probabilities of groundwater occurrence can often be made and the amount of slow and expensive exploration is greatly reduced.

The limitations now lie on the committee tables of the public authorities and private contractors who must generate the energy, foresight and funds to use these new tools.

Priorities in developing countries. Much of the international aid for developing countries takes the form of technical survey missions, whose reports normally have to draw attention to the lack of hydrological and meteorological records. The history of development in many countries of Africa shows a curious tendency for the offer of further aid to be in the form of yet more surveys, rather than in the systematic correction of the data deficiencies already reported. The World Meteorological Organisation has striven determinedly to correct this balance and it is fortunate that the insistent demands for more and better air-route weather forecasting are steadily increasing our knowledge of the rainfall input and of the factors contributing to evaporation. The priority thus obtained for the training of junior and middle-grade staff to carry out the routine measurements is a most important contribution by the developing air transport industry.

Measurement of water flow in streams or rivers requires more costly initial installations, even if only for the skilled time needed to 'rate' a cross-section so that flow rates can be deduced from simple records of water depth. Substantial measuring weirs may be needed, requiring road access to site for heavy loads.

Drilling of boreholes is essential for the location of underground aquifers and for the routine recording of their water quality, depletion and recharge. Drilling is expensive, particularly in countries with few roads, but it offers the rewards of discovering ready-made natural water reservoirs, which can be drawn upon and if necessary artificially recharged, by structures costing only a small fraction of the price of a storage dam. The difference may often be a critical factor in decisions on the development of irrigation schemes.

In logical development, installation of measuring structures and organisation of the staff and supervision to operate the routines and to tabulate and report the data, should all follow without delay on the recommendation of a competent technical reconnaissance. Practical examples of such timely action in Africa are the long-standing co-operative readings of water levels of Lake Victoria and flows of the Upper Nile, agreed between the British and Egyptian governments, which prepared the way for the highly successful development of the Jinja Dam at the lake outlet; the lake level was raised to generate electric power for both Uganda and Kenya. Substantial installations, including a $\frac{3}{4}$-mile-long measuring weir across the Sabi River, set up in Rhodesia following the 1950 report by a major group of engineering consultants, have secured the essential hydrological information on which the present spectacular irrigation developments are proceeding. In contrast, lack of adequate rainfall records, combined with lack of attention to the scanty data available, led to the costly failure of the British Groundnut Scheme in Tanganyika.

40

This is a field in which international organisations can help to provide both the technical foresight and the means for developing countries to act on it. The 'World Bank' (International Bank for Reconstruction and Development) supports many such 'pre-investment studies' in preparation for major developments. Improvement in national networks for collection of hydrological data is one of the basic objectives of the present International Hydrological Decade (1965–74) and may well be its major contribution to the development of man's water resources.

Routines and research. Such investigations into precipitation, streamflow and groundwater are usually regarded as reconnaissance and practical investigation rather than research, because incomplete data of widely different standards of reliability have often to be bridged by experienced guesswork in the synthesis of hydrological information which leads to design. The routine collection of the data is also not research, but it is of even greater importance. It is the basis on which much of the infrastructure of society is designed and by which all research into the water relations of our environment must eventually be interpreted. There is, however, a wide field of research into the physical causes of precipitation and evaporation and much active research into improved methods of measurement, recording and interpretation of hydrological data. Watershed research, on the interactions of these factors, and their integrated effect on streamflow, is discussed in Chapter 5.

The sampling of rain and snow

The catch of raingauges. For rainfall the conventional raingauge is still the basis of measurement, although studies of air turbulence and gauge-exposure effects have applied some qualifications to the interpretation of the readings. It is amusing to follow the development of the conventional rubric of a 30° or '1 in 2' safe distance for objects surrounding a raingauge. Horton's (1919a, b) pioneer experiments comprised a single row of raingauges perpendicular to a single row of trees. By repeated quotation, usually without the original reference, it has become incorporated into so many textbooks as to be accepted without question. Geiger (1950) did show that for forest clearings in Europe a 30° angle gave rainfall about equal to that measured in the open outside the forest, but he also showed that a smaller clearing, with a 50° angle, gave more shelter and a catch 5% higher than that in the open. In the tropics McCulloch (1962) found a 45° angle to be large enough to avoid distortion of the catch. Where high windspeeds are frequent, the fitting of

41

shielding vanes to reduce eddy effects at the rim of the gauge has been shown to be important and the World Meteorological Organisation recommends the Alter pattern of shield in which the gauge is surrounded by a simple circular array of hinged vertical pendent metal strips. The Nipher screen, a solid shield, shaped like a bucket without a base, is also used in the USA, while in the USSR the Tretyakov shield has been adopted. This consists of metal strips similar to the WMO shield, but they are bent and fixed under the gauge. Fortunately the effects of turbulence all act to reduce the catch of a gauge, so that comparison is simple, the gauge with the highest catch being most satisfactory. In well-replicated comparisons the Tretyakov gauge has been found to catch as much summer rain and more snow than the other patterns and has therefore been adopted for the vast Russian hydrometeorological network.

In sites remote from habitation, storage gauges, with various devices to reduce evaporation loss, give some limited information, but for studies which relate rainfall to streamflow the readings must be at least daily, and preferably more frequent. Here the 'technical fallout' from space research has contributed a useful device. Solar cells, developed to charge the batteries of spacecraft, are now used to supply the power for automatic weather stations (Sumner 1963). Rainfall operates a simple tilting bucket mechanism by which records of occurrence, together with a time-marking, are punched on to tape, and successful runs of up to six months have been achieved. Where the weather stations can be reached by telephone lines such rainfall records are transmitted automatically to data centres, while alternative equipment, still under reliability trials in the UK, permits interrogation of the automatic station at convenient times from the data centre, and transfer of the recorded information. Automatic radio transmission has also been developed, but this is expensive and restricted to sites of exceptional difficulty of access.

All such complex devices can be put out of action by a falling branch, a rolling rock or an excess of curiosity by wildlife. Tropical environments appear to teem with hazards of this nature, from solitary wasps which build nests of cemented mud in the narrow funnels of raingauges, to elephants which inspect and discard equipment. Far more menacing, of course, are the less responsible human visitors whose interventions in the raingauge network range characteristically from simple theft to target practice with gun or spear.

In many developing countries the difficulty and expense of servicing delicate electronic equipment still exceed that of the training and the supervision of staff for manual observations. A compensating advantage is the ability to maintain high replication of observations and thorough routines for checking of tabulations. Precipitation varies so widely that reliability of performance of a large number of stations within a modest margin of error is of far greater importance than higher precision in a single measurement; claims for higher precision are not therefore the deciding factor when selecting equipment.

Similarly, although raingauges should in theory be randomly distributed, convenience of access is of such dominant practical importance for reliability and continuity of record that it becomes a major deciding factor in location of gauges. However, as raingauge networks are established in uninhabited areas, often hill ranges important for water catchment, the freedom of choice conferred by the remote-recording and automatic devices gives new emphasis to the problems of rainfall sampling. The general influences of topography on rainfall distribution are a matter of elementary geography and everyday experience, but the methods of obtaining the best estimate of total precipitation over an area of pronounced topographical relief with the minimum number of stations are not yet adequately studied. Even in flat country at high

Table 1. Variability of raingauge catches.

Standard error of mean for 24 gauges on 12·5 ha of mountain slope, as % mean rainfall. Where a recognisable trend occurs, a more accurate estimate can be obtained from the same number of raingauges by dividing the area into equal strata and sampling each stratum by two gauges. (McCulloch 1962)

	As a randomised pattern	As stratified in pairs by altitude
Annual rainfall		
(1800–2500 mm)	1·0	0·5
A storm of 13 mm	2·5	1·4
A storm of 74 mm	2·0	1·4

latitudes there is an erratic areal distribution of summer rainstorms, but annual or longer-term averages tend towards uniformity. In semi-arid tropical country, where water is the limiting factor in land use and measurement is therefore important, the summer rainfall effects are accentuated and an intensive pattern of gauges is essential for the estimation of water input over a catchment basin area. Grouping of the raingauges, for statistical analysis, either by proximity, in flat country,

43

or by stratification wherever a regular trend such as a hill slope is discernible, has been shown to reduce the error of estimate both in the USA (Wilm 1943) and in the tropics (McCulloch 1962). Table 1 gives some examples of the precision with which mean rainfall is estimated over watershed experiments currently in operation.

Rainfall intensity. The rate at which rain falls is an essential hydrological measurement, made by passing the catch of a raingauge through a device which inscribes a record on a clock-driven chart. Manual transcription is laborious; recording can be electronic, with analysis by computer, but the costs are high and advanced standards of servicing are required.

Measurements by radar. An entirely new principle of rainfall measurement has been developed in recent years by the use of radar. Radio waves of about 10 cm wavelength or less are reflected by raindrops. A scanning beam of selected wavelength and signal strength can penetrate rainstorms and show on the radar screen a pattern of both density and location of rainfall. When the beam is swept over an area which has a sparse pattern of raingauges, photographs of the radar display give a valuable guide to the interpretation of the raingauge data, while a series of storm studies can indicate desirable locations for raingauges in relation to topographical features (Rockney 1960). The range for such purposes is up to 200 km. However, the use of radar as a direct measurement of area rainfall is not yet reliable enough to replace raingauges. Development of this method appears to offer the only practical prospect of estimating rainfall quantities over rugged terrain under conditions of high windspeed. Hamilton (1954) in the USA and Fourcade (1942) in South Africa have shown that in such circumstances raingauges cannot be expected to give a quantitative estimate.

Assessment of snowfall. A similar trend away from the physical catching of precipitation samples is even more essential in measurement of snow. The effects of air turbulence near to the gauge are very much greater on snow than on rain. In the USSR where spring snowmelt causes flooding of the major rivers, and has required major programmes of dam construction, the assessment of snow accumulation has been given much attention. Comparisons of snowgauges with various patterns of shielding have been assessed against measurements of snowfall in sheltered clearings in a uniform canopy of dense low shrubs. These probably afford the best available measure of snowfall and accumulation.

44

The experiments, carried out at the Valdai Experiment Station, some 300 km from Moscow, showed that the Tretyakov gauge and shield caught only 60% of the snowfall, while the Nipher shield, the WMO-pattern shield and other alternatives all caught less. For hydrological purposes, snow is therefore measured mainly by survey along 'snowcourses', or transects along which the depth and density of the snow is sampled at intervals. The irregular distribution and depth of snow and the highly variable density of the snowpack, as the result of successive partial melting and freezing, constitute a severe problem of sampling. A promising line of study, also in the USSR, is the direct assessment of the water content of the snowpack by its effect in attenuating the natural radio-activity emitted by the ground surface. Transects are metered immediately prior to the season of snowfall by registering the gamma radiation at successive sampling points. Field instruments already developed commercially for the search for radio-active minerals are adapted for use. The measurements are then repeated above the snowpack and the differences are related directly to water content. The method is now under test at many stations throughout the USSR, but is likely to be limited in practice to areas with adequate levels of 'background' radio-activity.

Streamflow measurements

Routines and problems. The methods of streamflow measurement in practical use vary widely in both cost and accuracy. For the layman they have an apparent simplicity which can be deceptive. By far the greatest output of flow data from rivers and streams is from daily readings, on fixed gauging posts, of the depth of water at points at which the cross-section of the channel has been surveyed. These sections are 'rated' by measuring the velocity of flow, at successive depths at a number of points in the cross-section. This exercise is repeated over the range of river levels from low water to high water. Where more accuracy is needed the depth of water is recorded continuously by autographic tracing of the rise and fall of a float. This is essential for streams which vary rapidly, but the interpretation still depends on the accuracy of the rating process. Streambeds are usually irregular in shape and the flow of water is complicated by turbulence and viscosity, so that such ratings are neither easy nor very accurate. The flatter and calmer sections of the streambed must therefore be selected for gauging sites, in order to reduce turbulence to a minimum, but unfortunately these are the sections in which deposition of sediment is at a maximum. Such sediment is usually deposited in irregular banks and shoals so that the cross-section of the streambed can be changed by a single storm when soil conditions in the watershed are unstable. Construction of a weir or barrage, restricting the flow to a designed geometrical shape, is a satisfactory but expensive solution, in which deposition of sediment may still be a problem.

Rating by radio-isotopes. Recent research has offered an encouraging alternative. This permits the selection of stable sites where the flow is over hard rock and where turbulence is too high for deposition of sediment. Stream discharge rates at successive water levels are then rated by dilution measurements using a dye or a chemical tracer. Dyes are partly absorbed on the streambed and banks, while chemical tracers such as sodium dichromate had at first to be used in unduly large quantities. Improved techniques, which can now detect minute traces (10^{-9}g) are now both useful and safe. The development of radio-isotope tracing techniques now offers yet another procedure. Radio-active chromium, iodine or colloidal gold are measurable at extreme dilutions and are not absorbed. They are reliably measured in quantities small enough to avoid all health hazards to down-stream water users (Plate 6). In some developing countries, where spate flows and high sediment loads present major gauging difficulties, these methods of rating stable streambed sections are particularly appropriate. Chemical dilution methods have been used with advantage by French hydrologists in West Africa (references in bibliography by Rodier 1963), and are used to check weir calibrations in many countries. Radio-isotope dilution methods have been tested successfully in Rhodesia (Ward and Wurzel 1968) and in the USA. Difficulties are likely to arise in all countries, however, from public fears of radio-active substances in water supplies at any dilution whatever, and these promising technical methods are unlikely to be generally adopted.

Weirs and flumes. For the smaller streams, whose measurement is often essential for the management of watersheds, the flow is usually led through special constructions whose rating is incorporated into the design. The most usual method is to halt the flow by impounding the stream behind a small dam or weir and then to measure the depth of flow as the water crosses a sharp crest of geometrical design and falls freely. The apparent simplicity of such devices can mislead the unwary, since the background theory is empirical, many formulae have been proposed for the rating curve, and there are a number of critical design requirements (King 1939). Such measuring structures require favourable

46

Plate 6. Dilution gauging with radio-isotopes.

Dilution gauging, either with chemical tracers such as sodium dichromate or sodium iodide, or with radio-active chromium, permits measurement of fast-flowing streams at stable rocky sites free from sedimentation.

topography. In steep mountain streams it may be difficult to secure, at a reasonable cost, a large enough stilling pool to quench the high velocity of the approaching stream. In flat country it is difficult to find a sufficient fall in levels and a suitable rock foundation to prevent water from seeping under or past the weir.

When spate-flow is characteristic and debris would frequently obstruct a sharp-crested weir, a convenient alternative is offered by standing-wave flumes, which are channels which constrict the streamflow into a wave whose height is related to the flow rate. In the western USA the Rocky Mountain Forest and Range Experiment Station has developed a pre-fabricated trapezoidal flume which can be carried in convenient sections and assembled in the field with a minimum of expensive skilled labour (Alden and Brown 1965). In Africa, this design has been adapted to the needs of developing countries where skilled labour is

Plate 7. Pre-fabricated measuring weir for experimental hydrology.
Built of acrylic resin reinforced with glass fibre, the pre-cast shell is bolted to simple angle-iron supports. The only field construction is then rough concrete foundations, but the weir must be accurately levelled. A pre-cast stilling-pool and metal weir-plate measure low flows. The approach channel was being cleared and straightened when the picture was taken.

48

scarce as well as expensive, by pre-casting the whole flume in a light plastic (acrylic resin reinforced with fibre-glass as used for boat hulls) which can be bolted to a series of simple angle-iron frames, reducing the field construction to concrete foundations of elementary form. The low flows need a more accurate device, and by pre-casting a small stilling pool in the form of a light plastic box, out of which the water falls through a small V-notch, the site requirements are further simplified (Plate 7; Gear 1968).

A heavier form of pre-fabricated measuring device recently developed for exploratory hydrology is a water-filled flexible plastic structure. This may well have applications in watershed experiments.

Even where designs are tested and pre-fabricated equipment is used it is necessary for workers in other disciplines such as forestry, meteorology, agriculture or physics, who may need to make streamflow measurements, to consult a civil engineer specialising in hydraulics about the design, siting and installation of such structures.

Flow records of large rivers cannot usually be taken in floods which overtop the measuring structures and prevent access by observers. The information is, however, essential for flood control design. A technique developed in the USSR for measuring flow rates of flooded rivers uses dye-covered floats which are dropped in line across the river from a low-flying light aircraft. Their progress is then followed by time-lapse photography from the same aircraft. Where a good topographic map is available, this can afford a useful estimate of flow. For measurement of large rivers in remote areas, a ground party is landed by boat or helicopter to plot a cross-section of the channel, and periodic flow measurements are then made from the air.

Analytical studies of streamflow records. Engineers commissioned to design storage reservoirs and flood-control structures by authorities which have not had the foresight to arrange preliminary study, have no opportunity to undertake experiments involving several years. There has therefore been an elaborate development of methods for the extension of scanty records by probability mathematics and by comparison with the most similar watersheds for which better records are available. Where there is a minimum of good records to show the shape of the stormflow from some heavy rainstorms, a number of empirical methods have been devised for extending this information to guess the response from the full range of storms. The best-known method is the 'unit hydrograph', which is an estimate of the pattern of stormflow response to rainstorms of various sizes, based on a characteristic pattern of response to a runoff of one inch depth over the catchment. This is assumed to hold for storms of larger amounts by increasing the vertical scale of the hydrograph. The method has given some useful approximations for large areas on the assumption that the land use is not changing. It is not of sufficient accuracy for use in watershed research where land-use changes are under study.

A routine calculation which is of direct interest for watershed research under land-use changes is the establishment of the 'base-flow recession curve'. During the dry seasons which are, either in summer or in winter, characteristic of many climates, the water stored in porous soil and rock strata on the slopes of valleys drains steadily underground to the stream. The flow derived from this movement of groundwater is known as the 'base flow'. In the absence of rain the rate of flow declines as the volume of stored water, and hence its hydraulic 'head', diminishes (as for water running out of a bath). This gives, for each valley, a reproducible curve for dwindling of streamflow with time. The shape of the curve is determined by the physical characteristics of the individual valley. Such predictable behaviour is obviously of importance for planning of water supplies and also, in watershed studies, for the seasonal return of the valley to a recognisable water-status.

Groundwater measurements

Conventional means of exploration. Attention has already been drawn on page 13 to the major resources of potable water and of capacity for its storage which lie in the aquifers or underground strata of sand, gravel, chalk or porous rock. These are as yet only partially explored even in well-developed countries, because exploration is difficult. Search can be directed into favourable areas by the normal process of geological survey, and can be further concentrated by specialist hydrogeological survey, using geophysical surface measurements such as electrical resistivity and shock-wave behaviour. Drilling is, however, essential to establish the quality and to make some estimate of the quantity of water available at various depths. In spite of recent advances in the speed and portability of drilling equipment the cost to water authorities remains substantial. Underground reserves have therefore been often ignored in favour of dam construction at many times the cost, but with apparently somewhat greater certainty of the outcome, in spite of the hazards of drought and sedimentation.

When potable underground water has been located, the rate at which it can be pumped for use depends both on the permeability of the aquifer and on the extent and reliability of recharge by water from elsewhere. Recharge may be directly from local rainfall or by complex seepage routes from rivers bringing water from distant rainfall. Ultimately groundwater recharge depends on rainfall and is therefore subject to interruption by drought, but the great volume of storage often affords a valuable continuity of supply in years when surface reservoirs are low.

The means in past general use for groundwater investigation were remarkably tedious and expensive, but recent developments of entirely new techniques hold great promise of accelerating the process of exploration. Until recently, direction and velocity of flow of underground water could be inferred only from the drilling and testing of a pattern of boreholes extensive enough for the plotting of contours of the free water surfaces. Draw-down of the water surface by pumping tests on individual wells is then used to observe the rates of recovery and also the effects on the water levels in neighbouring boreholes. The physics of this process suggests that it can only be highly approximate (Childs 1969). Permeability tests are also made on rock samples from the drilling cores. Contours of the underground water surface cannot be drawn effectively until most of the pattern has been drilled, while the necessity to site several boreholes in a group for pumping tests restricts their deployment in exploratory sampling over a large area. Costs of drilling vary widely with the nature of the ground, the depth needed and the accessibility of the sites; but substantial numbers of holes are needed, e.g. in a local exploratory study in Africa 50 reconnaissance boreholes were sunk to 65 m depth in an alluvial plain of 400 square miles (Hindson and Wurzel 1963).

Flow and direction from single boreholes using radio-active isotopes. In recently developed techniques, radio-active isotopes are used to measure both direction and rate of flow of groundwater across single boreholes, so that a small number of drillings can now be sited to better effect for the exploration of a large area. The tests are made more rapidly and the results appear to be at least as accurate as the conventional methods.

Radio-active substances have been tested intermittently for some thirty years as tracers of underground water, by releasing them in one borehole and seeking to detect them in others, but the method has several drawbacks and has not been widely used. In 1963, however, Mairhofer, working in Vienna in association with the International Atomic Energy Agency, established a method for measurement of the direction of flow in a single borehole. He released a solution of radio-active isotope at the level of the aquifer, to be carried across to the lee-ward side and adsorbed on the borehole wall and into the adjacent soil or rock. A scintillation counter housed in a lead shield with a window in one side is lowered on rigid rods and rotated to detect the position of highest radio-activity. Wurzel and Ward (1965) published an elegant simplifi-cation of the method for use in developing countries, by which the electronic equipment is left in the laboratory. The radio-isotope solution is released within a cylinder of iron gauze, carried by a rigid rod from the surface. The compass directions are marked on the gauze, which is re-turned to the laboratory for measurement of the distribution of radio-activity. The method has given good results for direction of flow over a range of aquifers from deep alluvial deposits to fissured crystalline rock.

For measurement of the velocity of flow, a known quantity of radio-active solution is released at the depth to be investigated and a counter, immersed in the solution, measures the fall in concentration as it is diluted by the flow of groundwater. This 'point dilution' principle was originally developed for use with chemical tracers, particularly in the USSR, but it did not then afford a convenient field technique. Practical field equipment for point dilution measurements with radio-isotopes was designed by Borowczyk, Mairhofer and Zuber in the Vienna laboratories of the International Atomic Energy Agency. This has needed some modification for field work in Africa, but it has been thoroughly tested in field conditions of the high-altitude tropics. Compact, portable tran-sistorised counting equipment carried in a suitcase is available com-mercially and for field work has replaced the former bulky and imposing array of boxes and panels (Mairhofer 1967).

A further new principle was applied to the measurement of under-ground water in 1958 by Buttlar and Went in New Mexico. They took advantage of the periodic occurrence of high concentrations of radio-active hydrogen (tritium, H^3) in the rainfall of the northern hemisphere, following the test explosions of atomic bombs in the upper atmosphere. The tritium has a known rate of decay giving a half-life of twelve years. The progress of underground water was traced by the peaks of radio-activity, and both rate of movement and length of storage could be calculated. The measurement of tritium involves rather elaborate and tedious laboratory methods of concentration, extraction and counting (Plate 2). It does, however, give useful results. In the field study of an alluvial plain in Africa already referred to, the results of all these methods, i.e. the conventional pump tests and hydraulic gradient cal-culations, the measurement of direction and rate of flow by using radio-isotopes in single wells, and the calculations from the tritium content of

52

a series of samples, have all agreed well within their estimated limits of accuracy (Wurzel and Ward 1967). More rapid progress in groundwater exploration can therefore be expected in the future.

Interpretation of results. Interpretation of field measurements is greatly assisted by two modern techniques in which fragmentary data are integrated into 'models' or systems proposed for the description of the functioning of underground water resources. The models are either sets of mathematical equations, tested in a digital computer, or sets of electrical circuits, in which the resistances to waterflow and the storage capacity of underground aquifers are represented by resistances and capacities of electrical components. Both techniques are in active use by water-resource laboratories.

Measurement of changes in soil-moisture storage

Every gardener is used to watering his plants at intervals and thus relying on their use of stored soil moisture for several days at a time. Similarly meadows, crops and forests may flourish for weeks after the end of a good rain season, where soils are both deep and porous enough to store and to yield adequate moisture. On a watershed scale, with deep-rooted trees, this storage can amount to 10 inches (0·25 m) or more over most of a valley and can thus play a very important part in water-resource calculations and management.

Soil samples are readily taken by the familiar soil auger (often a carpenter's wood-drill welded to an iron rod) for inspection and for chemical analysis; by oven drying it is very easy to find the proportion of water in soil by weight. To be useful in hydrology, however, we must know what volume this water represents, since we need to express it as a depth over a known area. Cutting of soil cores of known volume is in practice a rather tedious process although tools are available to do this in most soils. In very uniform soils, which are more characteristic of tropical than of temperate zone conditions, it is possible, by drying many samples of known volume, to establish a relation between the depth in the profile and the volume occupied by unit weight of soil. Samples can then be taken rapidly by a simple auger. Hand sampling is still useful for many problems of detail in agriculture and in horticulture but it has been rapidly superseded, for watershed work, by the 'slow neutron meter'. This is a far more costly instrument (about £1500) but it has the important advantage of 'non-destructive' operation, i.e. of sampling without soil disturbance. The same sampling holes are used repeatedly.

They are drilled through soil, gravel and loose rock and lined with steel or aluminium tubing to the depth of the vegetation or beyond it where possible. A 'probe' containing a radio-active source and a sensitive receiver for radiation, constructed as a robust field instrument, is lowered to successive depths. Radiation, originating as fast neutrons, is returned as slow neutrons only after collision with hydrogen nuclei in the soil; that of the water present greatly exceeds the hydrogen of the organic matter and soil minerals, so that changes in soil moisture are readily observed as changes in the electrical charges counted in a given time by the receiver. The counts are referred to a calibration curve which gives the volume of water per volume of soil. In addition to the somewhat elaborate nature of the equipment, the method has other complications, in that the radius of the sphere of soil from which the slow neutrons are returned varies with soil moisture content; the first 100 mm depth of surface soil is also inadequately measured. In spite of these difficulties the method is the best available for work on a watershed scale. Reliable and convenient designs of this equipment are now widely used in water-resource studies (Bell and McCulloch 1966, 1968).

In spite of the attractions of this modern electronic technique, it is well to remember that an immense amount of productive research has been accomplished by soil sampling and oven drying. In developing countries, where servicing of electronic equipment may involve long delays, many straightforward problems may be solved by soil sampling in less time than is needed to calibrate a neutron probe in local soil types.

Much valuable information on the penetration of seasonal rainfall into the soil and on its subsequent withdrawal by plant roots can be obtained from very simple electrical soil-moisture tensiometers, known as 'resistance units' or 'gypsum blocks'. In the original form, due to Bouyoucos and Mick (1940), these were two wires or wire meshes cast into a match-box-sized block of 'Plaster of Paris' (porous calcium sulphate) buried at various depths in the soil. The porous material is wetted and dried in equilibrium with the surrounding soil, while the electrical resistance changes characteristically from about 100 ohms when wet to some 10,000 ohms when dry. This is readily measured from electrical leads projecting from the soil surface. Over most of the useful range the logarithm of the resistance is directly proportional to the logarithm of the drainage tension. Practical examples of the use of this technique are given in Chapter 6. A serious limitation is that such resistance units are affected by soil salinity, especially by chlorides, so that they are sometimes not applicable in semi-arid country where their simplicity of operation is of most advantage.

54

Evaporation and transpiration

Evaporation rates are basic to hydrological planning, since every water-storage structure from a farm pond to a major reservoir which exposes an open water surface to sun and wind incurs an extra loss of water. Water use by vegetation is usually the major item in the difference between precipitation over a watershed and the streamflow from it. The combined flow of water as vapour both from direct evaporation and from transpiration is often referred to in the literature as 'evapotranspiration'. In hot dry climates this can be nearly 100% of the rainfall. No fully satisfactory direct measurement of this vapour loss over a landscape is yet possible; even the surface of a lake presents difficulties. Evaporation requires energy, which is received from direct solar radiation onto the land or water surface, from cooling of the vegetation or water mass and sometimes from advection or heat transfer from hot winds. Air movement also plays an important part by carrying away the water vapour and preventing saturation of the zone near to the evaporating surface.

Quantitative estimates. Under natural outdoor conditions it is not yet possible to claim that evaporation from lakes, rivers, marshes, fields and forests can be measured, but we can make some good estimates which, under favourable conditions, predict water losses effectively. These estimates are based on methods which fall into two main groups.

The first and oldest is direct measurement of water losses from evaporation 'pans' or tanks filled with water and exposed to sun and wind. A useful variant is a 'lysimeter' tank filled with wet soil in which vegetation is grown.

The second type of estimate is the calculation, from meteorological data, of the evaporation opportunities provided by the local atmospheric conditions and the rates of evaporation thus induced from vegetation freely supplied with water. These methods have arisen in the past two decades from studies of the physical components of the evaporation process. The resulting estimates are called 'potential evaporation rates'. They have proved the more useful and reliable for calculations of major water-development problems and are now in general use for water-resource management.

The literature of proposals for the estimation of evaporation from climatic variables is of formidable volume and variety. The formulae range from the simplest, by Khosla (1949) who used half of the mean air temperature, $T°C$,

$$\text{Evaporation, } E = \frac{T}{2} \text{ cm per month,}$$

55

to the complex empirical formula of Turc (1961) which includes radiation, air temperature and humidity. Linacre (1963), after reviewing a diverse array of such arbitrary formulae, concluded 'However convenient, they are less a short cut than a dead end.' Their origin, and the reason for their occasional further use, is that the data on which to base a more logical estimate often do not exist. Engineers designing water resources with only fragmentary data tend to use such formulae and to allow generous safety factors. Some empirical formulae have been widely used for particular purposes (e.g. Blaney and Criddle 1950, for irrigation; Turc 1961 and Olivier 1961, for reservoirs).

Formulae based on the Piche device, an inverted tube of water closed by a disk of filter-paper, have been widely used in France, but the device is less useful than the standard wet- and dry-bulb thermometers in the meteorological screen.

An equation which assembles routine meteorological observations on thermodynamic principles was first proposed by Penman (1948). The Penman equation provides the most general solution yet available and is currently the most widely used, although its application to the study of past records is limited by the rather infrequent occurrence of all of the measurements required (sunhours or radiation, air temperature and humidity, windspeed). The Penman equation is of direct service in river-basin studies of the water use of vegetation and is therefore described later.

Evaporation pans. Much of the early data in technically developed countries, and sometimes the only data for estimating water use in newly developing countries, is from evaporation pans; literature on this is voluminous and often contradictory and some notes on their main features therefore follow.

Practical but very crude approximations can be made from evaporation pans or tanks. A design used in the USA has been adopted by the World Meteorological Organisation as a world standard because of the very large numbers already in operation. It is the 'Weather Bureau Class A Pan', a shallow circular galvanised iron pan of 4 ft diameter standing on a gridded timber base. Its rate of water loss has to be adjusted by a correction factor to estimate the rate of loss from a lake and by further correction factors to estimate water use by vegetation. Direct sunshine on the sides causes the Class A Pan to grow far hotter than the water in a lake or reservoir, while the shallow layer of water is both warmed and cooled more rapidly. Direct comparisons with lakes or reservoirs are not at all simple.

56

In the most thorough set of studies yet undertaken, on Lakes Hefner and Mead in the USA (US Geological Survey 1952), the correction factors varied widely in the winter, and over the year ranged from 0·1 to 1·3 (Kohler, Nordensen and Fox 1955). The variations are due mainly to changes in heat storage in the lake. For very large bodies of water the

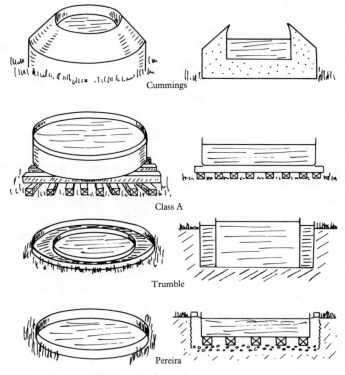

Fig. 4. **The insulation of evaporation pans.**

Evaporation rates from pans differ from those of lakes mainly because of their diurnal range of water temperatures. They can be made more efficient by insulation with rock wool, as in the Cummings design, with a water jacket, as in the Australian (Trumble) tank, or by an air-jacket as in the design used in the tropics by Pereira. In practice, these arbitrary devices are chosen for cheapness and standardisation, and the least efficient design, the Class A pan, is adopted as a world standard.

heat storage so dominates evaporation rates that the Northern Pacific and Atlantic Oceans have a half-year lag and evaporate more water in the winter than in the summer. The lake experiments, however, gave average summer values of 0·7 for the pan correction and this appears to have a general application in many countries. The factor is strongly influenced by the presence or absence of evaporation from the surrounding area. Ideally the pans should be surrounded by 'well watered short grass' of a sufficiently large area. In arid surroundings progressive increases in the diameter of a surrounding plot of watered grass appear to reduce the correction factor until it reaches about 0·7, after which further increases in the watered surround have no effect; no satisfactory experimental determinations have yet been reported as to the minimum area required, but a field is more appropriate than a small plot.

There are more efficient designs of evaporation pan in which the influences of the surroundings have less effect (Fig. 4). The pan may be insulated by thick walls (Cummings 1940) or more cheaply by sinking it into the soil within a water jacket (Australian Bureau of Meteorology 1961) or more cheaply still within an air jacket (Pereira 1959a). Since, however, all such pans are arbitrary devices needing a correction factor, standardisation is of more importance than efficiency. It is most important in routine practice to be able to inspect for leakages so that tanks set directly into contact with the soil are inconvenient for maintenance. A major difficulty in hot dry climates is to prevent birds and animals from drinking from the pan and to ensure prompt removal of very large numbers of flying insects which fall into it. In the tropics the author has even discovered a passing villager washing her laundry in this convenient pool. Dust settles to the bottom and alters the reflection of solar radiation from the galvanised iron floor, requiring periodic cleaning, which in dry dusty conditions may cause frequent interruption of daily readings. To overcome these tropical problems, in 1959 the water engineers of the six countries, then called Kenya, Uganda, Tanganyika, Northern and Southern Rhodesia and Nyasaland, agreed to use a standard one-inch mesh screen of thin-wire netting over the pans and to coat the interior with black bitumastic paint. These combined precautions have resulted in more consistent results which can be compared with greater confidence between stations. The evaporation pan is discounted by Sutton (1953) as a theoretically intractable means of estimation of open-water evaporation, but in remote areas a consistent run of such records is warmly welcomed by engineers, who all too often have to design storage works using evaporation estimates based on much less appropriate clues.

Evaporation from vegetation. For the research worker the hydrological effects are clearly presented in Penman's *Vegetation and Hydrology* (1963). For other members of a Water Board the following section offers some examples as a background to this much debated problem.

Water loss from crops, pastures or forest is even more difficult to measure than is the evaporation from open water. The simple and obvious method of growing plants in watertight containers and weighing them at intervals began with Van Helmont's classical experiments in 1652; similar observations have been made for some 300 years and have gained much information about plants but remarkably little about their transpiration. This is because the experiments were conducted on the plant and ignored its surroundings. Unfortunately the much more useful pioneer physical observations begun in 1688 by de la Hire, Court Meteorologist to Louis XIV, attracted little interest: de la Hire built lead-lined outdoor tanks $2\frac{1}{2}$ m deep and made comparisons of water use of grass and bare soil. Briggs and Shantz (1914, 1916) made most thorough and determined studies of the water use of plants in pots, both in the glasshouse and out of doors. They observed diurnal changes in transpiration and attempted comparisons with evaporimeters, but both plant physiologists and physicists were then still decades away from effective estimation of plant water use.

The transpiration ratio. Progress was delayed for perhaps half a century by an almost unanimous conviction of botanists that water use was an intrinsic property of the plant itself. This led to attempts to establish for each species a 'transpiration ratio' of the weight of water used to the weight of dry matter produced. The belief dies hard; and certainly in arid conditions where water supply is the critical limiting factor the ratio has some apparent constancy (Walter 1964).

Arkley (1964) in California salvaged much of the information compiled in search of the 'transpiration ratio' by restudying the experiments for which both soil fertility and atmospheric humidity data were available. He showed that where soil fertility was adequate 90% of the differences in transpiration ratio could be accounted for by climatic factors. Where climate had been fairly constant, soil fertility was able to account for 75% of the variability. It was the study of plant growth in lysimeters, by scientists of other disciplines, which finally broke through this mental barrier.

Lysimeters. Plants were grown in soil contained in tanks or 'lysimeters', from which drainage was collected, initially in order to study their nutritional gains and losses. The water quantities, however, led to erroneous estimates of water use until two critical requirements of the technique were understood.

Firstly the vegetation must be part of a continuous canopy, with no change in height or exposure at the tank edges. As already described for the evaporation pans the surroundings should be releasing water vapour

59

at similar rates; the 'fetch' of wind should be over a hectare or so rather than over a few square metres of similar vegetation. The foliage under test must not have extra ventilation, extra sunshine on the sides, or receive heat reflected from the tank rim or border. It is perhaps surprising that this has been often overlooked but the writer has seen an 'experiment' in which evaporation tanks bearing tall reeds from a swamp were sited on a concrete apron of an airport.

The second criterion is that the soil drainage conditions should be similar to the field conditions to which the estimate is to be applied. This requirement has caused so much confusion in the past that it is worth careful attention here. The case has been clearly stated by van Bavel (1961). A deep lysimeter can reproduce a soil drainage profile similar to that of a deep natural soil, but not identical with it if plant roots reach the bottom of the container where the drainage tension falls to zero. Wallihan (1940) published the first experimental demonstration of this effect. He compared two gravity-drained lysimeters with two others in which the drainage through porous plates was increased by suspended columns of 136 cm of water. The volume of percolating water was five times greater from the soils drained at the greater tension and oats grew taller and transpired more from the simply-drained tanks. Lysimeters designed to measure water use are therefore now drained through porous plates or pipes in order to apply the appropriate forces to oppose the withdrawal of water from the soil by plant roots. To achieve such drainage equilibrium it is most important that the structure of the soil pore-spaces should not be destroyed; the soil columns are therefore excavated as monoliths.

Exceptionally, where the maximum or 'potential transpiration' is measured, a water table is maintained within the lysimeter soil column so that the whole root system lies within the 'capillary zone' or zone into which water moves readily through continuous water films. There is thus no physical restriction on water uptake by the crop roots.

The best-known set of weighed lysimeters is probably the pioneer equipment at Coshocton, Ohio, where seven massive columns of 8 m² area and 2·5 m depth, in concrete boxes with an overall weight of some 60 tons each, have been weighed continuously for thirty years with a precision of 0·25 mm of water (Harrold and Driebelbis 1958). Major improvements in design have followed the conversion of weight changes into electrical signals by transducers, as in the intricately instrumented lysimeters at Tempe, Arizona (van Bavel and Myers 1962). Here weight changes and meteorological measurements are converted into computing code and punched onto eight-track paper tape for automatic analysis by computer. A very compact and robust weighing equipment designed in Australia by McIlroy and his colleagues at Aspendale is similarly recorded automatically for computer analysis (McIlroy and Sumner 1961; McIlroy and Angus 1963; Berwick and Sumner 1968). Weighing of very large soil columns, in which crops or even trees are grown, may be conveniently achieved hydraulically by floating the tank and registering the changes in the level of water in the supporting vessel. The Ouryvaev design, in which a central column is supported by floats in three

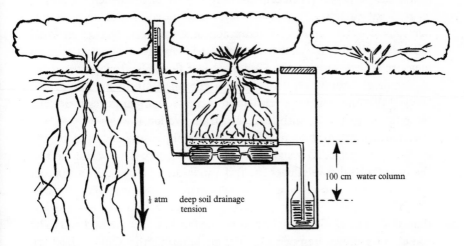

100 cm water column

¼ atm   deep soil drainage tension

Fig. 5. An hydraulic weighing lysimeter with applied drainage tension.

In a deep freely-drained tropical forest soil, laboratory study of undisturbed cores drained at ¼ atm tension (0·3 kg/cm² or nearly 5 lb/in²) showed moisture contents in close agreement with those found in the forest to a depth of 3 m (Pereira 1959b). Drainage of a monolithic block of soil, supported on a simple hydraulic weighing device, through a hanging column of water below a ceramic filter (Forsgate, Hosegood and McCulloch 1965) has been successfully used in study of a tea plantation (Dagg 1970). Most of the drainage occurs in response to the first 100 cm of tension and this is applied in the lysimeters.

surrounding tanks, is used in a range of sizes in many hydrometeorological stations of the USSR. A highly developed set of large 6 m diameter lysimeters at Davis, California, includes both mechanical and hydraulic weighing systems (Pruitt and McMillan 1962).

A simple but accurate design of lysimeter tank suitable for plant water-use studies in developing countries (Forsgate, Hosegood and McCulloch 1965) is in use in several parts of East Africa in studies of tea, sugar, pyrethrum, grass and cereal crops. A large undisturbed monolith of soil about 3 × 3 × 2 m is enclosed in a steel tank and drained under tension (Fig. 5). The tank rests on four flexible steel boxes filled with water all connected to a single water gauge which registers changes in weight equivalent to 0·25 mm of water. Dagg (1970) has reported very consistent results from this equipment in the tea crop, which agree well with measurements from an adjacent watershed experiment.

Transpiration tents. An attractive possibility of measuring transpiration directly in the field, by placing a transparent tent over small trees and shrubs, was developed from laboratory methods, initially by Decker and Wetsel (1957) and Decker, Gaylor and Cole (1962). The commercial availability of polyvinyl chloride or polyethylene sheet, which is durable, transparent to both long- and short-wave radiation, and almost impervious to water vapour, offered practical advantages. Decker enclosed individual shrubs in cylindrical enclosures supported by inflation, and measured the water vapour in a pumped air flow. No attempt was made to simulate natural ventilation and a thorough study by Lee (1966) showed that the restricted ventilation increased temperatures inside the tents by up to 10 °C, with increased transpiration losses.

A parallel attempt to use Decker's technique was initiated by the author in a brief FAO assignment to Israel in 1960. A powerful fan and the use of a baffled inlet over the full 3 m width of a cubical transparent enclosure permitted ventilation levels to match those measured in nearby shrub vegetation – while temperature variation within the tent was less than 1 °C. The air flow was heated by the fan compression and had to be cooled to ambient temperature before entering the tent (Schachori, Stanhill and Michaeli 1965). Comparative water losses from potted shrubs of pine, eucalyptus and cypress showed an average of 2·5% increase in transpiration inside the tent as compared with shrubs exposed in the open.

The technique was further developed and tested in irrigated plantations of bananas (Rosenzweig 1969), in comparison with water-use estimates from soil moisture changes, using a neutron probe. No significant difference could be detected in soil moisture withdrawal by banana plants inside and outside the tent, but the evapotranspiration as calculated from water vapour in the air stream was 1%, 8%, 10%, 2% higher than the estimates from the soil moisture changes over four separate periods of several days.

It is thus a possible technique, but is probably best used only under circumstances for which it was initiated, in which alternative methods are less accurate or more difficult to achieve.

Potometers. Simple and direct methods are preferable if they give the right answers; there has been much interest in estimating transpiration by measuring the rate at which a cut shoot or branch takes up water. The method is called 'potometry' although this term has also been used in the USA for weight loss measurements of plants growing in pots. Measurements by severing plant stems under water and connecting them to a metered water supply have persisted for over fifty years. The reasons for the failure were finally explained by ingenious experiments at the São Paulo University of Brazil (Rawitscher and Rawitscher 1949). By cutting transparent windows in plant stems the authors observed the immediate formation of numerous small bubbles in the xylem channels of the stem when this was cut under water; thus the sap flow is interrupted by the surface tension effects. Henrici (1943) devoted admirable energy to the very rapid weighing of cut branches of trees and bushes in South Africa, but this method also failed for the same reason.

Plant water stress. Although transpiration loss cannot be measured on detached leaves or branches the state of internal water stress of the plant can be readily measured. This is done by enclosing a detached leaf in a pressure chamber with the cut end of the leaf stalk protruding through a seal. Increase of gas pressure to balance the internal stress drives the liquid surface of sap in the cut stalk back to its original position. Its arrival at the cut surface is readily observed. The gas pressure gauge then gives direct information of the internal water stress of the plant (Scholander *et al.* 1965). The method is particularly useful in studies of the optimum irrigation for tree plantations since the equipment is robust and portable. Even when well irrigated, trees in hot weather can develop substantial water deficits in their foliage (Goode 1968). Millar and his colleagues (1971) have shown that even in a barley crop on soil wetted to field capacity the leaf water content is directly proportional to the vapour pressure of water in the surrounding air.

Potential evapotranspiration. In spite of the increasing sophistication and accuracy of lysimeters, however, it was the simplest possible design which, in the hands of a physicist in Britain and of a geographer in the USA, enabled them to demonstrate the dominance of climatic factors in plant transpiration. Using columns of infilled soil in which water tables were maintained, at constant level, from supply tanks, Thornthwaite (1948) at the Johns Hopkins University and Penman (1948) at Rothamsted published, independently, evidence that a continuous canopy of actively growing green vegetation, whose roots are freely supplied with water, transpires at rates controlled by the climate and not by the plants. Total water use is referred to by Thornthwaite,

63

but not by Penman, as 'potential evapotranspiration' and this term now occurs widely in the literature.

Thornthwaite was concerned mainly with the establishment of a basis for classification of climates; he used only the simplest of meteorological records, the monthly means of air temperatures. By fitting constants to data from North America and correcting both for season and for day-length (using a latitude factor), he was able to predict crop water use from his lysimeters. A further important aspect of Thornthwaite's work on climate classification was the employment of a water budget, of the use and recharge of soil moisture in order to estimate the length of growing season available. His method has been used effectively for study of irrigation needs in North America, e.g. Sanderson (1950), and quite arbitrarily by geographers for describing, without verification, climates well beyond the range of the original data. Since the fitting of constants is an arbitrary process Thornthwaite's results applied best within the area from which he assembled his data. Extrapolation to other areas has been less successful, and for tropical use the process would need re-calibration onto local lysimeter data.

Penman presented a quantitative combination, in terms of routine climatic observations, of the two main physical processes involved: firstly, the supply of energy for the evaporation of water and, secondly, the uptake and removal of water vapour by the atmosphere. Penman's expression for water vapour transfer was founded on Dalton's classical pioneer work of 1802. The energy budget estimates incoming solar energy, from radiation measurements or from sun hours, and its distribution partly by reflection, partly by outgoing long-wave radiation, partly as latent heat of evaporation of water and partly as direct heat transfer to the soil and air. Less than 1% is used in building up plant material by photosynthesis. Since available climatic data are limited, their use to describe a complex process requires various approximations. Penman's equation uses only air temperature and humidity, wind-run and sun-hours (or, if available, radiation measurements). It is therefore sometimes described as 'semi-empirical' although it is based on logical physical arguments. It has been tested over a wide range of climates with remarkable success. Stern and Fitzpatrick (1965) obtained good results with it in the dry monsoon climate of Australia. In high-altitude tropics it has given good agreement with experimental basin measurements (Pereira et al. 1962; Blackie 1971; see Chapter 6) and with pan evaporation (Dagg and Blackie 1970). From current technical literature it is more widely used than any of the many alternative formulae which have been proposed.

64

Radiation measurements. The main limitation on all climatic methods for estimation of evaporation is the difficulty of measuring the incoming solar energy which drives the hydrological cycle as described in Chapter 1. Accurate measurement requires exactly constructed electrical thermopile instruments whose very small output needs rather elaborate and delicate electronic recording. This is quite feasible at an observatory or main laboratory but is difficult in remote mountainous catchment areas and field irrigation schemes. Recording of sunshine hours is easier and Penman uses this approximation because it is widely available, but the radiation value of a sunshine hour varies with latitude and to some extent with climate (Glover and McCulloch 1958).

New instruments developed for the totalling of daily radiation are based on the silicon cells developed for charging the batteries of space-craft. These offer a new possibility of measurement of radiation when connected to a mercury-thread electrolytic integrator (Federer and Tanner 1965). Compact transistorised counters suitable for the field are now available for integrating the daily output of electrical sensors and thus replace expensive and delicate recording instruments of the potentiometer or bridge type. The electrical sensors have themselves been greatly improved and their construction made less expensive (Monteith 1959a; Funk 1959; Fritschen 1963).

The most frequently needed radiation measurement is that of the difference between the incoming and the reflected radiant energy, known as the 'net radiation', i.e. the radiation absorbed by the surface under study. By placing pairs of horizontal sensing surfaces back to back, facing upwards and downwards respectively, and opposing their electrical responses, the net radiation can be metered directly. The glass covers which protect the sensing elements on the standard Moll and Eppley designs of radiation instruments also restrict their response to that of the short wavelengths only. Renewable polythene windows, transparent to a very wide range of wavelengths, are used in the net radiation instruments.

Radiation estimates in tropical climates. A simple field device, by which daily short-wave solar radiation is estimated by evaporating water in an enclosed container and condensing it into a simple measuring tube, was invented in Italy by Bellani (1836). It was redesigned in modern materials by Gunn, Kirk and Waterhouse (1945) and adapted for field use in tropical agriculture by Pereira (1959a). There the radiation supply is so dominant in the evaporation process that the distillation device by itself accounts for between 85% and 93% of the variability in pan evaporation and avoids most of the difficulties of operation of open-water pans in hot climates.

Near the equator the instrument gives best results when the receiver is shielded by insertion into a small pit in the ground. In sub-tropical latitudes with bigger seasonal soil temperature variation a ventilated shield is used above ground. Both forms of shield give a very high correlation between the distillation readings and the radiation measurements of a standard thermopile (McCulloch and Wangati 1967). In high latitudes an earlier form of the distillation instrument has been used with a metal-coated glass receiver fully exposed to radiation reflected from the ground and from surrounding objects (Courvoisier and Wierzejewski 1954). Stanhill (1965a) compared the accurate standard thermopile with four simpler methods of estimating radiation under hot desert conditions in Israel. He found the Gunn–Bellani instrument to give the highest correlation with the thermopile, accounting for 98% of the variability in monthly radiation totals, 96% in weekly and 86% in daily totals. It is particularly appropriate where observers have little training. Where observers are well trained the approximation to radiation by a daily estimate of cloud cover was found by Stanhill to be simpler and cheaper and to give results of acceptable accuracy for irrigation under these extreme conditions.

Monteith and Szeicz (1960) tested the Gunn–Bellani radiation integrator in the tropical mounting under UK conditions at Rothamsted. It proved to be insensitive to the very low levels of radiation of the UK winter climate, but to be effective in summer conditions.

There is an increasing everyday use of evapotranspiration measurements in both temperate and tropical climates as a routine of efficient irrigation management. These estimates now play an essential part in the watershed studies as described in Chapter 5, where radiation exchanges are discussed in more detail.

Short cuts. In warm dry tropical conditions, where the ability of the air to accept water vapour rarely limits evaporation, Walker (1956) showed that the back radiation and vapour transport terms of the Penman equation approximately cancelled out and that the radiation term alone gave an estimate of useful accuracy. In the contrasting wet conditions of the cool humid climate of an English winter the vapour transport term alone, needing only wet and dry bulb temperatures and an average windspeed, can be used with a set of regional coefficients, to give potential evaporation in Great Britain for six months of the year (Smith 1967).

Mapping of potential evaporation. The detailed water-use studies of the East African regional laboratories have been extended to map the Penman evaporation estimate over Kenya, Uganda and Tanzania (Dagg, Woodhead and Rijks 1970). There the largest factor in the Penman estimate is the short-wave radiation. Although there are some radiation stations and many records of sun-hours, very large areas have no energy measurements at all. Similarly, there are few good records of windspeeds. Civil airports, however, record both cloud amount and Beaufort-scale wind estimates and Penman used these as estimates for some UK conditions. From the cloud cover a sun-hour estimate and hence a radiation approximation can be deduced. From the Beaufort scale observations a wind-run estimate for the vapour-transport term can be calculated. Even when using both these substitutions Woodhead (1970) reports the error of the mean annual potential evaporation to be only 15% while the error for the monthly estimates is 18%. Such rough estimates are of genuine value to the hydrological engineer, who must decide on the allowance to be made for annual water loss by evaporation when designing a reservoir. From the more accurate estimates at the well-equipped stations it is clear that potential evaporation varies little from year to year, in any one area, over the wide range of East African climates, so that an average of three years of data at a site should determine the long-term average within about 5%.

Advection. No routine method of computation from atmospheric measurements can yet account successfully for evaporation in strongly advective 'oasis' conditions in which a substantial proportion of the heat energy reaching the crop foliage is brought by hot winds, as when irrigated areas are surrounded by desert. Lemon, Glaser and Satterwhite (1957) have measured substantial effects on irrigated cotton in Texas. Hudson (1963, 1964, 1965) and A'Hafeez and Hudson (1965) have illustrated with very simple devices the range of conditions existing within a major irrigation scheme. They showed sharp changes in evaporation rates from the windward to the leeward edges of irrigated areas of the Gezeira scheme in the Sudan. Rijks (1971) further investigated advection in the Sudan Gezeira. Using sets of electrical resistance wet and dry bulb thermometers on three portable masts, he investigated the evaporation rates at successive distances from the upwind edge of an irrigated cotton field of 4·2 ha. He found evaporation over the whole field to be at 1·5 times the rate which could be sustained by solar radiation, the extra energy coming from hot, dry winds from the desert. Near the upwind edge of the irrigated field local advection effects increased the evaporation rate to 1·8 times that from the net radiant energy supply. Stanhill (1965b) has measured the effect in a tall maize crop in Israel in hot, dry surroundings and by plotting results from the literature has shown that the effect on evaporation is sharply felt for the first 50 m of wind 'fetch' over an irrigated crop, but decreases until after 200 m there is no evidence of further change. At present we have clear evidence that the oasis effect exists but no established means of calculating it from atmospheric measurements. Evaporation pans can give a crude indication and Fitzpatrick (1968) has mapped for Australia a seasonal index of advective strength based on the ratio between Penman estimates and pan evaporation. Accurate lysimeter studies with extensive watered surrounds showed evaporation from well watered grass to exceed that of the estimates by 20% annually, the energy being 20% in excess of that available from net radiation. A formula proposed by McIlroy goes some way to prediction of evaporation rates in these circumstances when calibrated by lysimeters, but these remain the only effective method of direct measurement (McIlroy and Angus 1964).

Aerodynamic methods. Lysimeters are, however, too costly and elaborate for replication on an adequate scale. Much research effort has therefore been devoted to an alternative principle, of measurement by aerodynamic methods, in which the upward flow of water vapour is estimated from measurements of air movement, temperature and vapour pressure at two or more levels. Thornthwaite and Holzman (1939) made an early attempt at such measurements in the USA, but their instruments were too crude for the purpose. Pasquill (1949) at Cambridge established that the method is possible. House, Rider and Tugwell (1960), using instruments linked to a computer, measured radiation balance, heatflux from the soil and evaporation by the aerodynamic method. They achieved an energy balance over a day to within 1%, but their instrumentation was not developed for continuous use. Penman and Long (1960) showed that in stable conditions the evaporation rates obtained from a wheat crop by the vapour flux method were some 10% higher than the calculations by the Penman formula for short grass, a result in accord with expectations from the roughness of the crop. Soumi and Tanner (1958) made similar measurements over an alfalfa crop which agreed well with a floating lysimeter. Rapidly responding instruments and eddy correlation techniques employing elaborate electronic integration equipment have been developed in Australia (Dyer 1961) and an assembly of this equipment, known as a 'dynatron' (Dyer and Maher 1965) has been used to study evaporation from marshes (Linacre *et al.* 1970). A very thoroughly instrumented study of the water balance and energy balance of a grain crop has recently been reported from the Central Plains of the USA (Hanks, Allen and Gardner 1971). In a widely-spaced irrigated sorghum crop in hot dry weather they found water losses to be highest in an area extending some 40 m into the field from the upwind edge of the crop. Over the first 100 m the advected energy drawn from the wind traversing the crop accounted for about 30% of the energy of evapotranspiration. In an adjacent sorghum crop without irrigation, which covered only 40% of the soil surface, they found the reverse effect. This plot was cooled by the wind, which carried off energy from it.

McIlroy (1971) reports progress on a simplification towards continuous recording from sensors placed close above the soil or vegetation. The instrument includes a bridge balanced by a servo-motor and much development may be needed to achieve a form rugged enough for long periods of field use.

In Chapter 5 progress is described in the development of an aerodynamic method of direct measurement of evaporation loss from above a forest canopy by mounting arrays of instruments on very tall masts. At present all such aerodynamic studies are directed to fuller understanding of the mechanisms of water loss rather than to the achievement of routine methods of measurement. *There is no early prospect of the development of a practical routine method and we still have no effective alternative to the lysimeter in advective 'oasis' situations.*

Recording of land-use changes

Responsibility for recording of land use and water quality. When water-resource authorities have taken in hand the assessment of rain and snow, of streamflow, of groundwater and of evaporation losses, they will usually find that to interpret such data, and particularly to interpret the trends of change with time, ancillary information is needed on land-use and water-quality changes. Those responsible for planning future water developments should be recording this extra information now.

Later chapters will summarise the important effects which arise directly from changes in the use of land. Such effects have had spectacular results where unplanned changes from forests or savannah wildlands to agriculture have increased the floods on mighty rivers such as the Mississippi and the Volga; it is not always realised, however, that important hydrological changes may result from deliberate changes *within* agriculture, such as from pasture to ploughed land or from over-grazed rangeland to improved pastures, or the reverse. Irrigation may

69

affect water resources even more directly, especially where problems of salinity occur, but hydrologists usually find the assembly of quantitative data on past irrigation practice to be very difficult indeed. Even in highly developed countries, records of land use and water quality have had little attention, and existing information is scattered among many public and private sources. An example from the USA is instructive. Some twelve years ago an investigation team from the US Agricultural Research Service collected precipitation data from the US Weather Bureau and streamflow data from the US Geological Survey, for some 700 valleys in which there were 30 years of good continuous record. Data for over 300 of these valleys could not be interpreted because there were no records extant of the chronology of the land-use changes. Trade statistics of produce, and agricultural records of crop acreages are reported by administrative districts unrelated to river-basin boundaries.

Aerial photography has greatly simplified this task. Over the past three decades this has made possible many effective studies in developing countries in which no other forms of record exist.

Study and annotation of aerial photographs is, however, easier and cheaper if made currently, rather than by reconstruction of the land-use information sometime in the future under the urgent timetable of a development scheme. Newly developing techniques of automated cartography, by which information from air photographs is coded on to computer tape and then automatically plotted on to maps, will greatly facilitate such recording of land use (Bickmore 1972). Baker and Dill (1971) for the USA, and Brunt (1961) and Bawden (1967) for Africa, give examples of land-use study from aerial photography. Photographic survey developed rapidly after 1947, as war-time techniques and aircraft were deployed for civil purposes and already repeated photography has yielded valuable evidence and estimate of land-use changes. Rapid development of colour photography has improved aerial survey of soils and vegetation. Scanning techniques using wavelengths in the far infra-red can detect surface temperature differences of 0·5 °C. Subsurface water has been detected by the cooling effects of the water use by vegetation. These techniques are developing rapidly and promise some very practical hydrological applications, especially in the detection of water pollution by industrial effluents (Grimes and Hubbard 1972).

Water quality needs more laborious routines of sampling and of chemical and biological analysis. We can be grateful to the exacting requirements of the steam locomotive for water of high quality at many railway stations for a great deal of our earlier records on which time-trends can be based. Sampling regularly at points where flow is also measured gives an early indication of any serious deterioration in agricultural efficiency by an increase in transport of soil in suspension. Decrease in irrigation efficiency, and the dangers of increasing salinity, may be detected in drainage water long before costly crop failures result. Similarly, early warning of urban and industrial pollution comes first from the laboratory reports of regular sampling.

It behoves water-resource authorities to accept the responsibility to assemble such information now and to keep it up to date since they will certainly need it in the future.

4

Recorded experience of the effects of forests on watersheds

When new land is settled water engineers and meteorologists are rarely on hand to measure the effects which the changes in land use will have on water resources, i.e. on the volume and pattern of streamflow and on the depth and quality of the water in storage underground. Forests are felled or burned and natural prairie or savanna grasslands are ploughed without reference to catchment areas. Even where some streamflow and rainfall measurements are established the records are usually held by different public departments, none of which keep records of the uses to which land is put year by year. The assembly of information from scattered departmental records requires much determination, good co-operation and even some luck. The FAO Working Group on the Influence of Man on the Hydrological Cycle sent the author in search of such data. With the help of National IHD Committees, particularly those of the USA, USSR and Australia, some good examples were found. They were summarised and widely distributed in a report to the Mid-Decade Conference of the International Hydrological Decade (Pereira 1969) which was sent by UNESCO to the National Committees of all the seventy-two nations taking part in the Decade. Some of the examples are quoted in this and in later chapters.

The effects of forests on weather

For a century or more, foresters have battled to protect their trees in the field against axe, fire and livestock, and in the Council chamber against endless threats of excisions of forest land. Many of them adopted and encouraged a mystical belief that the presence of the forest actually causes rainfall and hence improves streamflow. Many such opinions are on record, but is there any recorded evidence of such effects?

Forests and mists. There are indeed specific conditions in which

72

forests can influence the occurrence of mists. The condensation of mist on to foliage can cause impressively heavy drip from trees. Parsons (1960) recorded about 10 inches, during each of four rainless summers, of drip from a pine tree on the Berkeley Hills overlooking San Francisco Bay. In Japan belts of trees are planted along the coast to intercept the inland drift of sea fogs (Hori 1953; Matsui 1956). Under freezing conditions such interception is increased and is reported in Australia to increase catchment yield (Costin and Wimbush 1961; Costin 1967). In another specific climatic environment, where the clearing of tall forest occurred in an area having a hot dry season, the author found a rare example of raingauges in the right places to provide circumstantial evidence of the effect of forest on mist. On the Mufindi Escarpment, a tea-growing area in Tanzania, the clearing of some 300 square miles of tall forest for subsistence farming appeared to have halved, over the following ten years, the number of occasions on which slight rainfall had been recorded; there was no perceptible change in the annual total rainfall.

Snow-trapping by forests. Data from Russia have often been quoted to support the effects of forest on rain: with the co-operation of the National Committee of the USSR for the IHD, the author was able to discuss this evidence with the specialists concerned. The data from routine records of streamflow from a vast area of the north-eastern zone of the European USSR do indeed provide evidence, on a massive scale, that under a rather unique set of climatic conditions forests in the area give a higher water yield than the farmlands cleared from the same forest. The special conditions are that almost all of the annual precipitation occurs as snow and therefore that the major runoff events in each year are due to snowmelt. Rakhmanov (1962) presents an analysis of the records of thirty-five river basins ranging in size from 150 km^2 to 2000 km^2, all in the forested ecological zone of the European area of the USSR. Rakhmanov avoided the usual problem of the selective occurrence of forest in the highest rainfall areas by choosing watersheds whose whole area lay under an ecological climax cover of forest and in which the differences have been brought about by clearing of large areas for agriculture. He used data for fifteen basins in, the watershed of the eastern tributaries of the Volga River and from seventeen basins in the upper reaches of the Dnieper River, all lying in plains of low topographical relief. Over the four unusually dry years from 1936 to 1940 in which any transpiration effects of forests might be expected to be at a maximum, he found that as the proportion of forest cover rose from 10% to over

73

80% there was a strong positive trend of increasing annual water yield which rose from 3·0 to 5·8 m³/sec/km² and from 4·5 to 7·5 m³/sec/km² in the two groups respectively.

A parallel study over a much wider range of climates was made by Bochkov (1959), who studied over 100 river basins in the wide precipitation range from semi-arid country under 350 mm (c. 14 in) per annum to the forest zone of north-eastern European Russia with 650 mm per annum (c. 26 in). The basins had areas of from 50 to 1000 km² with a few larger examples in the dry areas. The results show both more forest and more streamflow from the wetter areas but the geographical range was large enough to relate both to latitude. Bochkov plotted both percentage forest cover and annual flow against latitude for the four classes of annual precipitation, 350, 450, 550 and 650 mm. He found that in the two middle zones the relation between forest cover and water yield is entirely accounted for by their relationships to latitude. In the driest zone no significant relationships at all were obtained. In the wettest zone, however, in which the climax vegetation is coniferous forest, Bochkov found no significant relationship between forest cover and latitude. Where the forest had been partially cleared for farmland he found an almost linear relationship between average annual flow and percentage forest cover.

Both Rakhmanov and Bochkov conclude that the extra water yield in the forested catchment is due to increased snow storage. Research into tree-felling in patterns which trap snow is discussed in Chapter 5.

Forests and rainfall. There is no corresponding evidence as to any effects of forests on the occurrence of rainfall. Bernard (1945, 1953), discussing the one million square kilometres of the Central Congo Basin, finds no support for any influence of forest on rainfall, but suggests that local forest clearing, by increasing the heat reflection, might introduce local instability and convergence, thus serving to promote rainfall in a very stable area.

The evaporation from forests has been alleged to increase rainfall; but McDonald (1962) illustrated quantitatively that evaporation from reservoirs and irrigated areas could not modify the dry climate of Arizona.

Penman (1963) has reviewed the evidence and the meteorological arguments concisely and concludes that although vegetation does affect the disposal of precipitation, there is no evidence that it can affect the amount of precipitation to be received.

The effects of fire on watersheds

The protective function of forest cover is demonstrated most dramatically when the covering from large areas is suddenly removed by fire. A forest fire not only destroys the trees but may consume much of the litter which protects the ground. Where watershed areas are being managed deliberately for the harvesting of water supplies, a major problem is their protection from forest fires. Some water supply authorities are reluctant to spend the very considerable annual sums necessary to maintain adequate fire-breaks and patrols.

An instructive example of the effects of fire on water supply is provided by the experience of the Snowy Mountains Hydro-Electric Authority in the Australian Alps. Here a major uncontrollable forest fire occurred in rugged inaccessible mountain country. The catchment areas of the Wallace's Creek and Yarrango Billy River, of 41 and 224 km^2 respectively, were burned out. The Authority had gauged these catchments for the previous eight years, with detailed sampling of suspended soil loads.

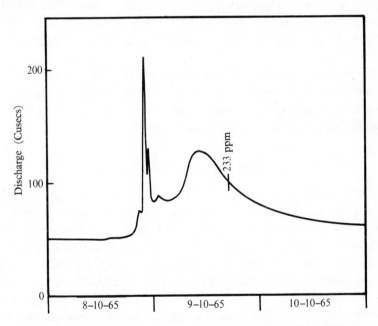

Fig. 6. **Spate flow following severe fire in the Snowy Mountains.**
The steep Yarrango Billy valley in the Snowy Mountains of Australia was under study for water yield and quality when the tree cover was destroyed by a severe fire. Suspended sediment increased one-hundredfold.

After the fire the flow pattern changed abruptly, with sharp flow peaks from the burned areas (Fig. 6). Rainstorms, which from previous records would have been expected to give rise to flows of 60 to 80 m³ per second (cumecs), produced a peak of 370 cumecs. The suspended sediment content at a flow of 60 to 80 cumecs has been increased by 100 times in comparison with the soil content before the fire. A storm occurring some seven months after the fire gave the highest sediment concentration yet recorded at Wallace's Creek. At a flow of 95 cumecs the concentration was 14·4% by weight, equivalent to 115,000 tons per day. This concentration is high by world standards. On the same day the Yarrango Billy River, with a flow of 47 cumecs, yielded an equivalent sediment load of 45,000 tons per day. The combined effect of the increased flow rate and the increased sediment concentration was estimated by the Authority to have given a total sediment load in Wallace's Creek of 1000 times greater than before the fire.

Similar damage was suffered in the San Dimas catchments of the Santa Gabriel Mountains in California when the dry chaparral scrub cover was severely damaged by fire (Rowe 1941, 1948). Watershed experiments in Arizona foothills under chaparral scrub, which were denuded by a very destructive wildfire, gave a tenfold increase in water yield (from 4% to 40% of a 600 mm rainfall). Erosion losses increased from one-thousandfold to three-thousandfold. The annual sand and gravel transport rose from 43 tons/km² in the previous three years to 50,000, 100,000 and 150,000 tons/km² respectively on three measured watersheds (Glendening, Pase and Ingebo 1961).

Beneficial use of water by forests

Where an evergreen forest uses water all the year round the clearing of the trees and the planting of short-season annual crops can result in a big reduction in water use. In Western Australia this land-use change has been occurring on a large scale for several years and the hydrological results are already embarrassing. Some 120,000 km² (30 million acres) of the native dry open woodland, under a winter rainfall regime of 400 to 600 mm annually, has been completely cleared of trees except for the borders of streambanks and drainage lines. The dominant woodland species were deep-rooted *Eucalyptus*. The land is now cropped in a rotation of wheat and short-season annual grasses and clovers. There are no perennial grasses capable of sustaining grazing and the sheep survive the dry season by eating the dead annual grasses and the seeds of the annual clover. With a hot dry summer giving an annual total for open-

76

water evaporation of some 2000 mm (80 in) or about four times the annual rainfall, it is indeed surprising that an excess of water should become apparent. Saline groundwater, however, lies beneath the woodland and where the trees are felled the excess winter rainfall raises the water level so that salt springs flow from the hillsides and spread over the low-lying areas (Bettany, Blackmore and Hingston 1964). Dead and dying trees along several hundred miles of drainage lines offer striking indications of the continuing progress of a land-use change which ignores hydrology.

Detailed surveys of the salted areas were carried out in 1955 (Burvill 1956) and again in 1962 (Lightfoot, Smith and Malcolm 1964). In the seven-year interval there was an increase of 720 km^2 (180,000 acres) in the salted land to give a current total of 1200 km^2 (300,000 acres) on a somewhat conservative basis of estimation. Even more important than the loss of land is the effect on the quality of the river water. Most of the streams were originally fresh, since under the woodland conditions the saline groundwater remained beneath a rather impervious layer of clay. There are unusually good records of the water quality since it was checked carefully by railway engineers from 1880 onwards. The Blackwood River, for example, was carefully checked for salinity in 1880 and was fresh enough for railway use. By 1910 there was heavy settlement and clearing for wheat production in the upper catchment areas of this river; by 1920 the water was already too salty for use in locomotives. Today it is too salty even for general irrigation and can be used only for salt-tolerant crops such as apple trees. Several water-supply dams which were constructed for the railway at the beginning of this century, to hold water of tested quality, have since been abandoned as a result of increasing salinity. The State Authorities have accepted the evidence as conclusive and have restricted clearing on some catchments in order to protect important sources of water. Beyond these catchments, however, more than a million acres (4000 km^2) are being cleared every year.

The problem appears to be one of finding deep-rooted perennial species of pastures or of economic crops. Salt-tolerant species of grasses and of edible shrubs have shown limited success in the saline areas but increased transpiration is needed for the restoration of the hydrological and salinity equilibrium of the countryside. In the meantime drainage works to contain and reduce the damage from saline outflows are becoming increasingly necessary to prevent the loss of more productive land. The problem is illustrated neatly in the watershed management for supply of the main storage reservoir for the capital city of Perth. In a series of drought years an attempt was made to increase the yield by ring barking to kill the trees in the upper watershed. The yield was indeed

77

improved but the excess water raised the level of the underground saline supplies while the loss of surface protection caused a sediment problem from erosion of the soil surface. As a result the Forest Department was requested to re-establish a tree cover by planting the upper catchment with pine. A plantation of 12 km² (3000 acres) was established. This has successfully corrected the difficulties and has been maintained to protect the water supply.

While the clearing of forest above a high water table can be expected to result in a rise in the water levels, an example of the reverse effect is worth recording. In the USSR under conditions of precipitation as snow on frozen soils, followed by spring snowmelt where infiltration rates are critical, a rise in the groundwater level has been observed as a result of planting trees. Large-scale plantings of broad shelter belts of mixed hardwood and conifer forest 50 to 100 m wide have been made over the past century while intensive planting of narrow shelter belts only 10 to 50 m wide and 200 to 300 m apart is continuing today. Observations on groundwater wells within and outside the older forest shelter belts have continued since 1892; they show after snowmelt a clearly defined rise in the level of the water table under each belt of forest over adjacent levels.

Forest plantations and streamflow in warm climates

Practical experience of the problems and controversies arising from the planting of trees in watershed areas is more extensive in South Africa than in other countries. Much of the high rainfall areas have a climax vegetation of evergreen forest, of which the dominant species are various olives (*Olea* spp.) and the tropical conifers (*Podocarpus* spp.). These forests have been largely destroyed by axe and fire. The vegetation, still dominated by fire, is now composed of grasses and scrub. The timber needs of a rapidly developing economy led to large-scale plantations of eucalypts, of pines and of acacia (grown extensively for wattle-bark for tanning). Poplars are also grown for matchwood. Wicht (1967a) pointed out that 'Forestry in the USA started in the forest; forestry in South Africa started in the veld.'

While the forest plantation was beginning on a small scale, an early pioneer of irrigation development (Gamble 1887) protested that

'It is a remark frequently made by the travellers that the climate of South Africa is becoming dryer. Springs which were once abundant are now weak, and rivers that formerly flowed almost constantly now seldom run. The hippopotamus used to be found in pools in the

78

Kuruman River in Bechuanaland, which are now quite dry except in an unusually wet season. Is this in consequence of less rain falling or from some other cause? The observations made for 47 years at the Royal Observatory and the scanty records elsewhere give no support to the view that the general rainfall is materially decreasing.'

The search for trends in rainfall records has continued with remarkable industry but with little success, in part because the character of random distribution has been often ignored. A genuinely random distribution of annual rainfall does not preclude long spells of successive dry years or of successive wet years. Such 'runs' of like occurrences can occur by pure chance, as can be demonstrated by tossing a coin (Kerrich 1950).

The experiences of South Africa have been shared by many other African countries and indeed they have been repeated throughout the warmer semi-arid regions of the world. Over a wide range of latitudes, in the absence of human interference, forested high-rainfall areas maintain perennial flows to lower and drier areas. In high latitudes the well-distributed rainfall usually maintains this process in spite of human settlement. In lower and warmer latitudes with more violent rainfall occurring in more concentrated rain seasons and separated by hot dry weather, the hydrological balance is more vulnerable to disturbance. With the increase of human and of cattle populations, of grazing, of trampling and of cultivating, together with, above all, the influence of more frequent fires, the characteristic result is a decrease in the ability of the soil surface to absorb heavy rainfall. This increases overland runoff and leads to spate flows in streams. The decreased infiltration fails to recharge underground storage aquifers which in turn fail to maintain springs and to supply dry-weather streamflow. For the same reason water levels fall so that boreholes and the shallower wells dry up. Increased pumping from rivers and boreholes to supply the growing needs of urban industry and agricultural irrigation draws even more heavily on the diminishing resources so that at each recurrence of drought the public concern quickens.

This continuing deterioration was evident enough in South Africa to incur a Select Committee in 1914, a Drought Investigation Commission in 1920 and a Desert Encroachment Committee in 1951. Between the last two enquiries the human population approximately doubled. The Desert Encroachment Committee was particularly concerned by reports of wind erosion and desiccation of soil laid bare by massive over-grazing.

As a result of this progressive desiccation, the forestry plantations came under heavy attack from an anxious farming community, sometimes in spite of the geographical context. By 1932 the possible adverse effects of forestry plantations were minimised by the prohibition of the planting of trees near springs and within 20 m of streambanks. The effects of this measure are considered in detail in Chapter 5. In 1935 the Empire Forestry Conference toured the country, discussed the effect of forest plantations on water supplies and recommended the setting up of research stations. The first of these at Jonkershoek, in rugged mountains some 30 miles from Cape Town, was set up immediately, in 1936; the second was opened in 1945 at Cathedral Peak in the steep grass-covered Drakensberg Range, and two more were added recently.

The development of the forest industry continued, as did the public controversy about its effect on the water supplies. The serious droughts of 1965 and 1966 sharpened the debate so that in 1967 a strong Inter-Departmental Committee on Afforestation and Water Supplies was set up to review the evidence. The Report (Malherbe 1968) discounted any suggestions of diminution of rainfall.

After a careful review of all available evidence the Committee found that afforestation as yet covers only 9–14% of the high-rainfall area and that 'no generally serious shortage of water can be confidently ascribed to it'. They recommended that where forestry is practised in areas of water scarcity the rotation should be as short as possible. They described forestry as making efficient and profitable use of water where it fell in the catchments, while transmission losses and inefficient furrow irrigation used water wastefully in the farming areas. The commercial return per cubic metre of water was calculated to be higher from forestry than from irrigated agriculture. Farmers were advised to make maximum use of government subsidies to build dams for the storing of surface water in the rainy season.

Where urban and industrial water supplies were needed for development and must come from the areas suitable for forestry, the Committee agreed, however, that afforestation should be restricted and that the catchments should be managed under the lightest vegetation cover which would give protection against soil erosion. They noted that the owner, if thus restricted from planting of trees, should have a claim for loss of income from the plantation.

Where forest plantations are grown for paper pulp the very great volumes of water required for the paper mill, which uses 150 tons of water for each ton of paper, can present a major problem; this is more a matter of pollution with chemical wastes than of evaporation loss.

Afforestation and water supplies in cool climates

As pointed out in Chapter 1, the provision of water supplies to meet the growing demand of urban and industrial development is already a serious problem for the UK and the USA. Particularly in the close confines of the British Isles, forestry is a conspicuous activity (Locke 1970) and effects on water supplies have been heavily challenged, often with more enthusiasm than science. Much surplus winter streamflow still goes to the sea, but already the development of major reservoirs, and the heavy demand on their capacity, has raised the need for the maximum water yield from their catchments.

The relatively low evaporation opportunity, characteristically exceeded by gentle and frequent rainfall for most of each year, renders both vegetation and soil less vulnerable to misuse than are the landscapes of hotter climates. The protective role of forest in catchments can often be acceptably replaced by a vigorous grass cover. Because the impending scarcity of water in Britain was not taken seriously as a national problem until the last decade, we do not yet have the records which would contribute useful information about the effects of forests, while research is newly begun. Research results are discussed in Chapter 5. They already amount to a reasonable certainty that even in cool moist climates forests use rather more water than pastures or cropland, and where water harvesting for the industries of major cities involves the investment of large sums of public money, the difference in water use can be significant. Where poor soils, distances from markets and other such limitations on the profitability of farming make forest plantation crops the most economic use of the land, the costs and water yield of alternative methods of land management become critically important. Abandoned land in this climatic zone usually reverts to scrub woodland whose water use is likely to be little less than that of a productive forest plantation.

Reafforestation of eroded watersheds

The world's outstanding example of the large-scale rehabilitation and development of the land and water resources of a river catchment system is that of the Tennessee Valley (Fig. 7). Close study was made of the records of one small valley of 40 ha (88 acres). This was natural woodland which had been part felled for timber, with patches cleared and cultivated and the rest overgrazed, frequently burned and reduced

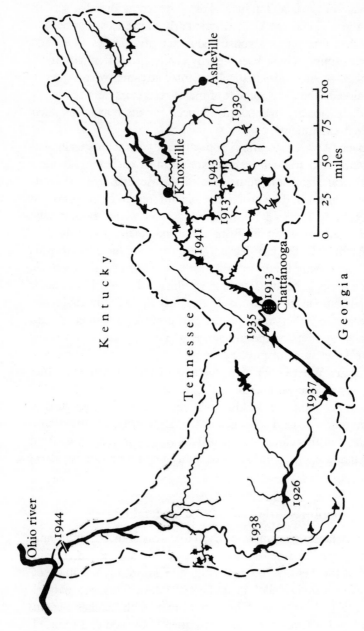

Fig. 7. **Water-resource development in the Tennessee Valley.**

Founded upon an initial development for hydro-electric power in an impoverished agricultural region, the Tennessee Valley Authority has integrated power, water-transport and irrigation projects over some 90,000 km² of watershed development. Since all of these enterprises are dependent on improved land use for controlled supplies of clean water, soil conservation has been a major objective of TVA policy, achieved through rising standards of agriculture and forestry.

by misuse to a state of active soil erosion. This description, from the records of the Tennessee Valley Authority (TVA 1962), is typical of the damage done by subsistence agriculture in many lands under many climates (e.g. Chapter 7). The stream gauging began in 1941 and the years 1941–5 were recorded for 'calibration' to characterise the behaviour of the stream in the eroded state of the valley.

In 1946 restoration began. The surface runoff was reorganised along gently sloping routes by means of 'cutoff drains' and contour furrows; the whole area was planted to pines (Loblolly pine, Pitch pine, Longleaf pine and Slash pine). Erosion gulleys were planted with Black Locust. In all 100,000 trees were planted, with good success. Hydrologically, the results were dramatic. The floods and soil erosion were eliminated, but the trees used half of the water flow. Peak rates of torrent flow were reduced by 90% while the sediment load of the stream was reduced by 96%. Thus muddy torrents were replaced by half the quantity of clean water as a controlled flow.

Molchanov (1960) has summarised a century of observations and opinions published by the numerous Russian writers on the subject of forests on water resources. He concludes (Chapter XIV) that for maximum water yield forests should be distributed in the form of shelter belts on the contour. As a general result, 6% of the watershed area under contoured forest strips halves the overland runoff of a completely agricultural watershed: 30% to 40% of the area thus afforested will ensure that the entire surface runoff is transferred to subsoil and erosion is thereby prevented.

Balance of advantage. Surveying this scattered experience, it is clear that forests should neither be felled nor planted on a large scale without a study of the potential hydrological changes which may be expected. The research described in Chapters 5 and 6 already enables confident qualitative predictions to be made, together with some approximation to quantity. Large-scale changes in land use can often be foreseen over many years, particularly when major water-harvesting projects are in prospect. There is good sense, both practically and scientifically, in the early initiation of pilot schemes to provide measurements. The reduction in guesswork can save very large sums of money.

5

Research on forested watersheds

Streamflow comparisons

The main data available for hydrological study are the records of stream-flow measured by water engineers in their assessment of local and regional resources. Many years of intensive research on these data by mathematical analysts have been devoted to studies of flood and drought, and have led to great advances in river control, in which the digital computer is now a routine tool. It has rarely been possible, however, to relate these studies of river flows to specific changes in the land use and management of the watersheds, although deteriorations in streamflow regimes have often been associated with land misuse. *The following discussion will be confined to research in which the change in land use has been a specific part of the experiment.*

For forest hydrology the evidence is being won from two types of watershed experiments, firstly those of long-term comparisons (regressions) between streamflows, and secondly those of intensive studies in which an energy balance is applied to the annual water budget.

The first method assumes that the variabilities of topography, geology, climate and vegetation interact in too complex a fashion for detailed analysis and that the only integrating measure is that of the flow of the stream itself. Weather variability from year to year imposes the necessity for long-term observations of streamflow both before and after changes in land use. Most of our established evidence in the subject has come from this method.

Historically, as noted in Chapter 1, foresters have been in resident charge of streamsource areas and they began early to measure the water yield of their forested valleys. In Europe there are records dating from 1867 in Moravia, now part of Czechoslovakia (Némec, Pasák and Zelený 1967). There the valleys were gauged as the result of a controversy about the hydrological influence of forests and two such gauges, established later in 1928, are still in operation today. In Switzerland Dr Hans

84

Burger attracted worldwide attention by his thorough studies of the flows from two similar valleys with contrasting 30% and 99% proportions of forest cover; these have been gauged since 1903 (Engler 1919; Burger 1943). They gave useful qualitative indications of flow characteristics. Burger concluded that streamflow from the agricultural valley was greatest under heavy rainfall and least after periods of drought. In spite of the long records, however, the uncertainty of estimation of rainfall (from gauges outside the catchments) and the geological uncertainties of possible leakage through the valley floors, have prevented any satisfactory quantitative conclusions about the water use of the vegetation (Penman 1958).

Clear-felling experiments. The first experimental study in which a planned land-use change was carried out in order to observe the effects on streamflow, began in the USA at Wagon Wheel Gap, Colorado, in 1910 (Bates and Henry 1928). Here streamflows from two similar watersheds of about 200 acres each were compared for eight years. One valley was then clear-felled and records continued. From the calibration period the flow of one stream could be predicted from the other by simple regression calculations. After the clear-felling the annual yield of water was increased 17% above that predicted from the flows of the unchanged control valley. The differences were observed to diminish rapidly as sprouting aspen restored the vegetative cover.

One-third of the total land area of mainland USA is under forest; this area receives more than half of the total precipitation and produces three-quarters of the total streamflow of the country. The rapid development of the USA brought many conflicts of interest in water resources and the effects of forests were assessed by streamflow comparisons over a wide range of environments. In 1934 the US Forest Service set up the world's largest outdoor hydrological laboratory, at Coweeta in the Appalachian Mountains, after an intensive geological search for sites in which the valleys could be expected to be watertight. The experimental area of 1740 ha has thirty small basins each producing a continuous stream. The annual precipitation is about 2000 mm; only 40 mm occurs as snow (Hursh, Hoover and Fletcher 1942). In spite of the similarity of Coweeta's forested watersheds, it is difficult to find groups of valleys of matching size, shape, slope and orientation, so that the usual scientific procedure of replication of trials is severely limited. Reinhart (1958), exceptionally, did find a set of comparable valleys supplying water to New York, and achieved some useful replication.

Encouraging results were obtained with calibration periods of a decade or less (Hoover 1944; Wilm 1949) on the assumption that annual rainfall totals are normally distributed (Kovner and Evans 1954).

Multiple-valley experiments with forest plantations. Wicht (1943) in South Africa has brought the streamflow comparison method to its highest degree of development by replication in time over a set of six watersheds. Sited in rugged mountain territory at the Jonkershoek Research Station, these experiments were begun in 1940 for study of management policies for steep scrub-covered mountain watersheds. The main treatments are controlled burning and replacement of scrub with Patula pine plantations. By planting at eight-year intervals on a forty-year forest rotation, comparisons of the effects of the pines are made at different stages of tree growth. Comparisons among such a set of valleys overcome some of the difficulties of applying the mathematics of probability to the prediction of flow of one valley from the measurements of flow from another (Wicht 1967b).

An even more intractable difficulty arises from the differences in the range of weather extremes experienced during any two periods of valley comparison when each period covers several years. The grouping of wet years and of dry years to give major departures from long-term averages of rainfall is a pronounced regional feature of river-flow records (Hidore 1963). That such departures can occur by random chance (Kerrich 1950) has already been noted in Chapter 4.

Although the prescriptions for mathematical analysis are indeed hard to satisfy in watershed studies, the reality of the results of one such simple study was triumphantly demonstrated at the Coweeta Hydrological Laboratory (Hibbert 1967). Here two small watersheds were compared for only three years of preliminary calibration; thirty-six monthly comparisons of water yield were made before one watershed was clear-felled in 1940. The immediate result, in the first year after cutting, was an increase in annual streamflow equivalent to 373 mm depth over the watershed area (14·6 in). The hardwood forest was allowed to regrow, and the streamflow decreased until, 23 years later, the flow was only some 75 mm greater than that predicted from the full forest cover. The cut was then repeated in 1962: the increase in streamflow was almost identical with that produced in 1940. Fig. 8 shows the annual 'deviations from regression', i.e. differences of observed flows from the yields predicted by the relationship to the control watershed.

The curve, which fits the figures well, shows that the decline of yield-increase is approximately proportional to the logarithm of the time in years since the felling of the forest (Kovner 1956).

86

Summary of comparisons of streamflow. At the International Symposium on Forest Hydrology in 1966, Hibbert surveyed the total published data from catchment experiments on changes in forest cover. He found results from thirty-nine experiments, ranging from Eastern to Western USA and including data from Japan, East Africa and South Africa. Most of the experiments were of the simple paired-valley type which he neatly described as 'calibrate, cut and publish'. The immense

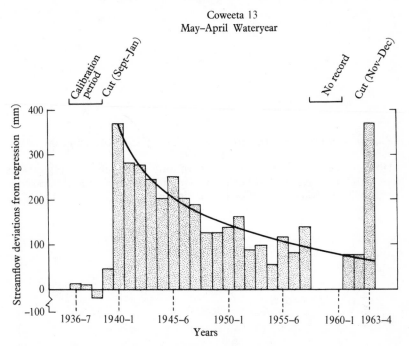

Fig. 8. **Increase of water yield after clear-felling a forest: a unique confirmation.**

After only three years of comparison of the monthly water yields of two small forested watersheds, one was clear-felled in 1940. Water yield increased by 373 mm depth over the watershed (14·6 in). As the forest regrew, yields declined logarithmically with time. Twenty-three years later, the cut was repeated. The yield increase was almost identically repeated. (Diagram from Hibbert 1967.)

scatter of the streamflow response in the first year after felling is illustrated in Fig. 9a. The one consistent factor is that a streamflow increase of 450 mm of water per annum over the cut area is a maximum under all of the wide range of environments tested. Most of the forest fellings produced less than half of this response. At Coweeta there were sufficient experiments under comparable conditions to observe that felling of

87

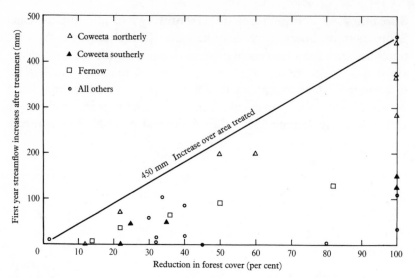

Fig. 9(*a*). **Summary of experiments on water yields following the felling of forests.**

Hibbert (1967) summarised the results of thirty-nine published experiments from both the USA and elsewhere. The only consistent factor in the responses was that all lay within the limit of 450 mm increase in yield over the area felled.

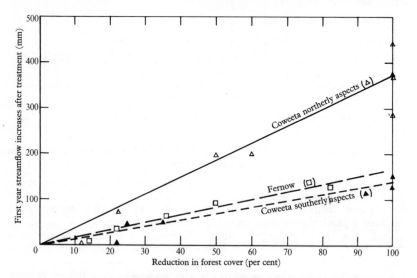

Fig. 9(*b*). **The effect of the aspect of watershed slopes on the water-yield response to clear-felling.**

Hibbert (1967) demonstrated in this diagram that, among the many watersheds of the Coweeta and Fernow Experiment Forests, those most effectively exposed to sunshine lost more water from the forest litter and undergrowth after clear-felling and thus showed less gain in water yield than those on north-facing slopes.

88

forest on northerly and north-easterly slopes produced about 350 mm of extra streamflow from the area cut, but that much smaller increases were obtained from clearing southerly slopes (Fig. 9b). In spite of this variability, world evidence is strong enough to conclude that in well-watered areas, streamflow response is proportional to the reduction in forest cover (Hibbert 1967).

Prediction of the water yield from planned watershed management practices is certainly the most urgent of the estimates needed in water-resource development, but these valley-comparison experiments can predict only in relative terms. We are thus very little closer to the under-standing of watershed functions which is necessary for the extension of results from experimental valleys into other environments.

At a recent conference on experimental watersheds, held in New Zealand in December 1970 as part of the programme of the International Hydrological Decade, there was a general conclusion that the era of 'calibrate, cut and publish' is now over and that a far more detailed study of components of the hydrological cycle is necessary.

Balancing the budgets for both water and energy

Most of the difference between precipitation and streamflow is due to the evaporation of water from plants and from free water surfaces of lakes and rivers. The next major advance in water-resources assessment came from the direct estimation of these vapour losses. Measurement of the water use of plants has been reviewed in Chapter 3. To recapitulate briefly, the direct estimation of plant water use as a component of the watershed budget became possible in 1948 when both Penman in Britain and Thornthwaite in the USA published evidence that when water is in free supply, the evaporation and transpiration from a complete canopy of green vegetation can be predicted directly from climatic factors. Thornthwaite (1948) used only the most widely available data of air temperature and day-length, and established an empirical relationship for North American data. Penman (1948) combined a quantitative budget of the incoming solar radiation, the heat used for evaporation, the back radiation and the heating of the surroundings, with an estimate of the ability of the atmosphere to carry off the water vapour thus pro-duced. Most of the energy for evaporating water is supplied directly by short-wave solar radiation, and where water is in free supply, evaporation, including transpiration, disposes of a large part of this energy. Penman first compared the estimates of evaporation from an open-water surface (E_0) with water use (E_t) by a canopy of short green grass, which in terms of modern agricultural production is still one of Britain's major crops. Four seasonal ratios were established at Rothamsted:

		E_t/E_0
Spring	March–April	0·7
Summer	May–August	0·8
Autumn	September–October	0·7
Winter	November–February	0·6

The first application of these estimates to watershed data was encouraging. Penman (1950a) calculated E_t values from long-term meteorological records of 100 stations, and showed that they accounted effectively for the long-term difference between the records of rainfall and river flow for all 40 of England's catchment areas from which long-term data are available. These covered the British Isles adequately and varied in area from 3 to 4000 square miles. The average annual evaporation loss over the British Isles is approximately 14 inches (355 mm), ranging from 13 inches in parts of Scotland to 24 inches in the southwest. The agreement of such completely independent estimates gave sound evidence that the results of assembling the meteorological data were real. Thus the average conditions over England's well-watered landscape approximate to a continuously green transpiring surface. A physical factor which favours such an estimate of evaporation opportunity over whole watersheds is that these conditions vary comparatively little over large areas.

Penman (1962) subsequently shortened the calculation of crop water use by adopting the findings of Monteith (1959b) that all ordinary crops reflect about one-quarter of the incoming short-wave radiation. Evaporation from the crop surface is therefore computed directly without reference to the rate for open water. Reflection values for forests are not yet well established because of the difficulty and expense of mounting instruments above the canopies. For watershed studies, in comparing different land uses, a reference level of evaporation from an open-water surface remains convenient, as is demonstrated in the following chapter.

Although such meteorological estimates may appear to be very approximate, their immediate importance to watershed research is that they are derived *independently* of the rainfall and streamflow estimates and therefore provide a valid check on the difference between them. We thus have the three major items in what can now be legitimately termed an approximate 'water balance', but there are two further important items.

Precipitation = Evaporation + Streamflow ± Storage ± Leakage.

At the end of each summer or dry season some of the water will have drained from soils and porous rock and some will remain in storage. Unusually wet or dry years may alter this residual amount substantially but in most years each river returns to its own rate of low flow, indicating that about the same depletion of storage occurs annually.

The importance of the 'base-flow recession curve' was referred to in Chapter 3. Analytical techniques for establishing the form of these base-flow recession curves are in the realm of the specialist hydrologist. The methods are empirical and amount essentially to recognising and putting together the fragments of the curve traced out by the streamflow in the intervals free from rainfall and overland flow.

The average date of minimum flow is usually selected as the end of the water year, at which an attempt is made to estimate storage and to calculate an annual water budget.

Seasonal soil-moisture storage changes. Where soils are deep and vegetation is deep-rooted a valley will continue to lose water by transpiration in a long dry season and will develop a substantial soil-moisture deficit. This deficit must be filled up from the next rains before effective recharge of deeper storage in underground porous rock aquifers can occur.

With the neutron probe, as described in Chapter 3, it is now possible to measure the seasonal soil-moisture deficits in the root range on a practical field scale, although the counting procedure is still rather slow when many soil profiles are to be recorded. Even where soils are deep, stone-free and fairly uniform, the installation of sampling tubes and the calibration of the meter constitute a time-consuming volume of work. Where tree roots penetrate deep into rock rubble the initial problems are severe. Calibration procedures involve reconstruction of rock rubble and soil assemblies of known water content. Schachori, Stanhill and Michaeli (1965) installed aluminium tubes, using a rock-drill, to depths of 10 m among limestone boulders and red clay on Mount Carmel in Israel. They showed that Aleppo pine (*Pinus halepensis*) dried out the profile to a depth of 5 m during the summer.

Seasonal groundwater storage changes. Where there is a water table near to the surface direct measurements of storage changes can be made from the rise and fall in well levels, but these have a deceptive appearance of simplicity. The surface of the groundwater is usually far from horizontal. Many wells or boreholes are needed to establish the underground contours of the water surface and to verify that these conform to, or at least are contained by, the surface boundaries of the catchment basin. This is, fortunately, often found to be true, but there are many examples of groundwater surfaces following the contours of old buried landforms rather than present surface topography. The measurement of a fall in water level cannot be used as a water quantity, because the outflow has been due to drainage of only the larger pore spaces of the layer of rock or soil vacated by the groundwater. This quantity is known as the 'specific yield' of water-bearing strata; it can be found very approximately by pumping tests. It can be measured accurately only by much laborious sampling and laboratory study of soil or rock specimens, carefully taken to preserve their pore spaces intact.

Estimates of groundwater storage changes over a catchment basin are therefore only approximate adjustments to the water balance even when conditions for measurement are favourable. In more difficult circumstances they may provide only 'quantitative clues'. Such end-of-season storage changes diminish in importance for cumulative studies of several years.

Leakage of watersheds. Leakage of watersheds is far more difficult to investigate. Leakage into a watershed can usually be avoided in experimental studies by siting them in the top of the headwaters of a river system so that there is no possibility of drainage into them from above. Leakage out of an experimental watershed cannot be directly measured. Experimental watersheds are therefore selected only where careful study of the geological structure of the area has indicated leakage to be unlikely. Gross leakages can be inferred where streamflow totals, evaporation estimates and storage-change estimates together fall substantially short of observed precipitation. With an independent energy budget to estimate evaporation, it becomes possible under favourable circumstances to test the water-balance estimates against a soil-moisture budget which can be verified by direct sampling. The deep stone-free soils and the sharply differentiated wet and dry seasons needed for this type of study occur geographically in the high-altitude tropics, and some results of experiments designed to use these advantages are discussed in Chapter 6.

When reporting watershed studies it is therefore essential to discuss the grounds on which leakage is either assessed or neglected. Reports of watershed studies in which the problem is not even mentioned should be regarded with reserve.

The physical principles of the combined water and energy budget are simple (Pereira 1959b). The first is that all of the water which is ob-

92

served to enter a catchment basin by precipitation must either be stored therein or must leave it as water vapour, water flow or seepage. The second is that the law of conservation of energy applies to a landscape as well as to a laboratory experiment.

Measurement of incoming energy

Radiation and reflection at the top of the forest canopy. The energy budget over a landscape concerns radiation in two ranges of wavelengths. The first, the direct sunlight, comes in the short wavelengths emitted by incandescent sources; this radiation is either visible or close to the visible range in the ultra-violet and the near infra-red. Such sunlight reaches any point in the watershed either directly from the sun, the 'solar beam', or indirectly by multiple reflection and refraction by the sky and clouds and reflection from surrounding rock, soil or vegetation (Fig. 10). The diffuse radiation from the full dome of a sky with much bright cloud can exceed that of the direct solar beam, although a more usual proportion would be about 20% of the total incoming radiation. When clouds interrupt the direct beam of the sun, the whole of the radiation is received by diffusion.

The second range of wavelengths comprises the radiant energy given out by any warm body as it cools to the temperature of its surroundings. Clouds, water vapour and carbon dioxide in the atmosphere absorb heat energy from short-wave radiation and emit it in long-wave radiation, some of which reaches the earth's surface. The surface of a watershed itself emits long-wave radiation to the sky. By day this is slight in proportion to the incoming short-wave radiation but by night the loss of the long-wave radiation cools vegetation and soils. Under heavy cloud cover the incoming and outgoing long-wave radiation are of similar amount so that the ground is usually cooled only very slowly. Under a clear night sky there is negligible incoming long-wave radiation to balance that lost: more rapid cooling occurs which may result in the formation of dew when the air near to the ground is moist or the occurrence of frost when the air is dry.

Incoming radiation of all wavelengths is partly reflected and partly absorbed. The ability of the surface to reflect radiation is called its *albedo* which is measured as the proportion of energy reflected. Differences in the land use on a watershed, e.g. forests, pasture, arable crops or bare soil, create differences in albedo which have important effects on the energy balance and hence on the water balance. The reflection ranges from 12% for a pine forest to 40% for desert. Clearly the

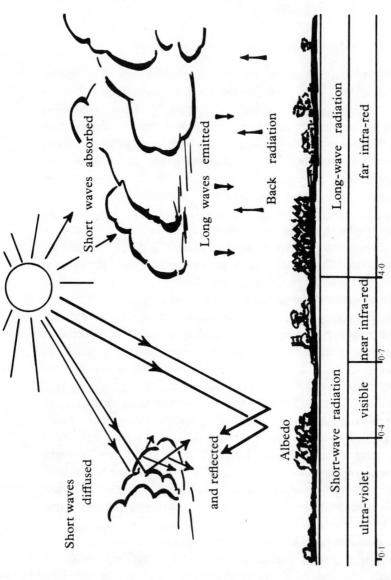

Fig. 10. **Radiation exchanges at a watershed surface.**

The land surface receives short-wave radiation both directly from the sun and by diffusion from the sky and clouds. Part is reflected and part is absorbed. Some of the absorbed heat energy is lost as long-wave back radiation.

lower the albedo the more energy is absorbed to become available for evaporation and transpiration or for heating of the surroundings. Ångström (1925) published albedo values as: grass 26%, oakwood 17·5%, pine forest 14%.

Incoming energy can be measured conveniently by instruments near to the ground, but the reflection and re-radiation of heat by a forest can be measured only above its canopy. A pioneer attempt to take such measurements was made in Italy by Pavari (1937). He suspended thermographs from captive balloons at heights of 500 m and 1000 m above pine plantations, oak scrub thicket and bare ground. He observed temperatures over the pine forest at 500 m to be cooler than those over bare ground or scrub. Geiger, in 1935, built towers to carry thermometers above a forest canopy (Geiger 1950). Stanhill improved on Pavari's balloons by mounting radiation-balance instruments on a helicopter. This created considerable instrumental difficulties, but the results were in accord with ground-based estimates of water loss by the various surfaces studied. The albedos thus measured are summarised in Table 2a. They agree well with those of Ångström. Similar values recorded from instruments mounted on aircraft in experimental flights over England and Wales are given in Table 2b. Tropical rain forests give similar results. Some recent data from Kenya, obtained by a mobile instrument tower (Blackie and Rawlings 1972) showed albedos of: tea bushes 20%, bamboo forest 16%, tall rain forest 9%.

Those who have occasion to fly in light aircraft over hot dry countries are well aware of the 'bumps' or turbulence due to heating of the air over desert or dry ground in contrast to the more stable conditions over forest or over open water.

Masts tall enough and strong enough to carry instruments above the canopies of high forest are very costly structures, but the logic of the physical studies of water use has brought research workers of different lands to the same solution. The use of towers, begun by Geiger in Germany before the Second World War, was followed by construction of elaborate equipment at the Valdai Experiment Station in the USSR (Federov 1957). Five steel masts 40 m high, hinged on massive bases and raised by powerful electric winches, were installed to carry the instruments above and within a pine forest 26 m high. The instruments were housed in small shelters suspended from cables strung between the towers. Krestovsky and Federov (1970), summarising the results of the detailed water and energy balance studies at Valdai, reported that evaporation from forested and open-field watersheds for a long period is approximately the same (about 500 mm per annum) although they are differently affected by summer and winter conditions. In wet years evaporation from forested watersheds is up to 12% greater than from fields, while in dry years evaporation from fields may be 10% greater than from forests.

In 1968 the Coweeta Hydrological Laboratory in the USA began an ambitious attempt to measure the radiation balance above a steep forested mountain watershed, by mounting radiation instruments to

Table 2. The albedo or reflection of solar
radiation by the land surface.

(a) Reflection by land surfaces in Israel	
Land surface	Albedo
Open water (lake)	11·3%
Pine forest	12·3%
Evergreen (Maquis) scrub	15·9%
Citrus orchard	16·8%
Open oak forest	17·6%
Rough grass hillside	20·3%
Desert	37·3%

Short-wave reflection as measured by instruments mounted on a helicopter by Stanhill, Hofstede and Kalma (1966).

(b) Albedo values from airborne measurements in England and Wales	
Agricultural grassland	24%
Deciduous woodland	18%
Towns	17%
Conifer plantation	16%
Heather moorland	15%
Peat and moss	12%

From a summary of evidence, presented and mapped by Barry and Chambers (1966).

monitor north and south aspects of slopes from high towers on the perimeter (L. W. Swift, Jr, report in preparation, 1972).

In Britain in 1970 the Institute of Hydrology has built two 32 m towers in a very large area of uniform pine plantation of some 200 km^2 with an average height of 15 m. These towers carry arrays of sensitive instruments for the study of wind speed, water vapour pressure and air temperature as well as radiation from sky and forest (Plate 8). The objective is to evolve methods of direct measurement of the upward flow of water vapour from the forest canopy and of the heat exchange of the forest with wind and sky (Stewart and Oliver 1970). Records of reflection of solar radiation by this very dense pine forest (spaced at 10 m^2 per tree), taken every minute from 5 m above the canopy and assembled by computer, showed marked seasonal variations (Stewart 1971). Albedo values averaged 8·9 in June and 11·7 in December. The all-year average of about 10% was substantially lower than that reported in Table 2b, which was presumably taken over less dense plantations.

Plate 8. Measurements of micro-climate above a forest canopy.
A study of the evaporative characteristics of tall vegetation at Thetford, Norfolk. From two instrument towers each 32 m high, windspeeds, temperatures and vapour pressures at eleven different levels are measured; incoming and reflected radiation are also recorded. (Institute of Hydrology.)

Orientation of catchments. Within a valley, common experience in high latitudes distinguishes sunny slopes and cold slopes. Crops which are particularly sensitive to radiation, such as grapes, have been grown throughout history in vineyards sited to receive maximum sunshine. Energy balances for complete watershed basins are, however, calculated from horizontal plan areas, using radiation measurements on horizontal surfaces, on the assumption that over a complete basin the variations cancel out.

The pioneer hydrologist, Horton (1932), proposed that the average orientation of a valley be described by an inclined plane fitted approximately to its rim. Lee (1964) developed a computer-programmed method for achieving the same approximation by multiple regression, using a series of points defined by co-ordinates in three dimensions and equally spaced around the watershed perimeter. Lee pointed out, however, that since the orientation of a watershed will have affected the history of weathering, and in consequence the distribution of soils and vegetation, any direct effect on the water balance of exposure to radiation may be masked by several variable factors. Contour mapping of mountainous terrain from aerial photographs already provides the topographical basis for analysis. Lee and Baumgartner (1966) presented a study of the distribution of slopes in 347 km² of Bavarian mountain forest land, but they did not have the means to relate the results to the water balance.

Swift and van Bavel (1961) made a direct test on four experimental watersheds of the Coweeta Hydrological Laboratory by calculating seasonal correction factors for the radiation received on the slopes. Three of the valleys gave annual patterns and totals similar to those of horizontal surfaces, while the fourth valley showed a substantial departure in winter, when water use is very low.

Corrections for orientation are thus at present a refinement which, hopefully, we shall need to apply when our techniques for measurement of the other variables are good enough to justify the effort. (See also Reifsnyder and Lull 1965.)

Studies of special forest components of the water balance

The possibility of a genuine water balance in which some quantitative estimate can be made for each item, concentrates more detailed attention on other facets of the water budget such as interception of light rainfalls, snow-trapping and snowmelt, and the special effects of streambank vegetation.

Interception of rainfall. Most people have at some time sheltered under trees in light rainfall and are aware that few raindrops reach the ground; also that it is unwise to lean against the tree trunks, down which water may be flowing. Such observations have led to a great deal of research on the interception of rainfall by different types of vegetation and its re-evaporation from the surfaces of leaves and branches. The work has attracted many short-term studies because of the apparent ease of measurement. In reality it involves formidable sampling problems, as

98

water concentrates into irregular drip-points below the canopy (Wicht 1941).

Enough data have already been published to establish that interception depends primarily on the physical character of the rainfall. Very light rainfalls of small drop size may be totally intercepted while heavy rainfalls of large drop size rapidly saturate the canopy, which thereafter transmits most of the water received. This is a physical process largely independent of the biological character of the foliage. It is well illustrated by Fig. 11 in which rainfall interception over a range of storm

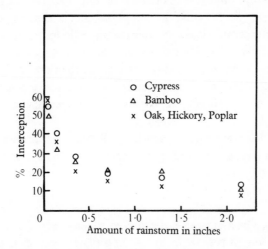

Fig. 11. **Interception of rainfall by tropical and temperate forests.**
The interception of raindrops by a forest canopy is a physical process governed more by the amount of the rainstorm than by the species of trees present. The curves compare detailed measurements for a natural bamboo forest and a plantation of cypress, both in Africa near to the Equator, with those from a forest of mixed hardwoods in the Appalachian Mountains of the USA.

sizes is compared for tropical bamboo forests, a cypress plantation near the equator in East Africa and a temperate-zone hardwood forest in the Appalachian Mountains (Pereira 1952). This comparison arose from a visit by Dr Charles Hursh, with data from the Coweeta Hydrological Laboratory in the USA, to the experiments in East Africa. The data are well fitted by hyperbolae, i.e. the curves are of a known shape. The same relationship to size of rainstorm has also been shown to apply to dense stands of Douglas firs in Oregon (Rothacher 1963). Thus the hydrological importance of interception by a continuous forest canopy depends directly on the pattern of rainstorms and is similar for many species of trees.

There are obviously great differences in the total surface area of the foliage of different species of trees, shrubs and grasses, with the greatest contrasts, on an annual basis, occurring between deciduous and evergreen species. It would clearly be pointless to attempt measurement of all the possible combinations of climate and vegetation. Fortunately sufficient detailed data have already been published to make possible a critical selection of future studies. Zinke (1967) surveyed the large volume of American evidence and showed that the summaries made by Horton (1919b) and by Kittredge (1948) are still valid. Zinke concluded that: (1) Interception loss is mainly a function of size of storm. (2) Most hardwood trees give similar interception effects during the growing season. (3) Interception loss is greater from needle-leaved trees than from broad-leaved trees. (4) Hydrologically interception represents an additional but transient form of water storage in the vegetation of a watershed. (5) The main interest now centres on the relative contributions of transpiration and of direct evaporation to the overall water loss of the wetted foliage.

Similar conclusions were reached by Russian experimenters over a wide range of tree species and ages. Their results have been summarised by Molchanov (1960).

In dry climates the effect of stemflow is of importance to trees by concentrating light rainfall within their root zones (Hoover 1953, Voight 1960). On a smaller scale the effect has been shown to be important for drought resistance in tall crop plants such as maize (Glover and Gwynne 1962).

For *comparisons* of streamflow the loss of water vapour from a watershed is the sum of evaporation and transpiration produced by a particular pattern of land use and there is no hydrological advantage in attempting to separate the measurements. For *prediction* of the possible effects on water yield of the planting of forests in a climate of known rainfall pattern, however, both the relationship of interception to storm size and some estimate of the direct evaporation are needed: the detailed mechanism is a necessary part of our understanding of the water use by vegetation.

In wet climates the evaporation opportunity is more important. If the evaporation of the water intercepted and held on the leaf surface merely replaced transpiration and occurred at the same rate there would be no need to enquire further, but three main factors can affect this rate. These are the energy available for evaporation, the ability of the air to remove the water vapour and the rate at which the roots can extract water from the soil, i.e. whether or not the transpiration is occurring freely at the full potential rate. Thus light summer rain falling on to a forest canopy with transpiration restricted by dry soils will greatly increase the rate of loss of water vapour. However, the same rate of light rainfall in calm misty conditions on a cold wet landscape, as in high-latitude winter conditions, will have little effect on the rate of vapour loss since neither heat energy to evaporate more water nor air movement to remove more vapour is available.

Rutter (1966a, 1967) and Leyton, Reynolds and Thompson (1967) have made measurements which indicate that, in the climate of the midlands of England, intermittent rainfall occurs frequently under conditions in which sufficient energy and air movements are available: evaporation of the films of water spread over the very large surface area of wet forest canopies then exceeds the transpiration rate. Research to discriminate between evaporation and transpiration from the foliage is very much more difficult than the measurement of through-fall. Although the methods used have as yet been rather crude (such as weighing of detached branches or comparisons with meteorological 'wetness recorders') there is little doubt that the effect is real. Comparison of contrasting physical paths of the water molecules passing through the plants with those evaporating from external films directly into the atmosphere does give good reason to expect that under a climate having frequent cycles of wetting and drying the interception of light rainfall will lead to a net hydrological loss (Rutter 1966b). The rate of evaporation then exceeds the rate of supply of solar energy, the extra energy apparently coming from the cooling of the air as it passes over the forest canopy.

Even in dry weather estimates of the aerodynamic 'roughness' suggest that a forest should, by the Penman analysis, transpire up to 10% more water than grassland. In rainfall regimes supplying much intermittent light rainfall this 10% difference may be increased to 20%. This could lead to reductions of 50 mm to 75 mm in runoff in some of England's more critical catchment areas although it would be of little consequence in the high-rainfall areas. In the lysimeter study of tea bushes in a climate of abundant rainfall, described in Chapter 3, Dagg (1970) found little effect of water loss by the frequent wetting of the foliage.

Rutter et al. (1971) have shown good agreement between calculations of water storage on the canopy of a pine plantation in the UK and trough measurements of through-fall. At present research is directed to explanations of the process as a prelude to more effective measurement; the mathematical physicist has a further contribution to make before

101

this effect can be calculated with confidence from the rainfall incidence and other routine meteorological data.

In contrast the lesser air turbulence over smooth short grass results in relatively little difference between the water loss from wetted grass or from freely transpiring dry green grass. However, as soon as a drying soil limits the transpiration of the grass, interception of light rainfall can again cause a net loss of water.

Interception and condensation of cloud, fog and mist. The drip and stemflow from trees surrounded by cloud, fog and mist are difficult to measure; the hydrological effect appears to be confined to the trees themselves. One of the better sets of measurements reported from con- ditions of abundant water supply was that by the Pineapple Research Institute in Hawaii (Ekern 1964). Here Norfolk Island Pines (*Araucaria heterophylla*) under conditions of abundant water supply, with annual rainfall of 2600 mm (104 in), condensed an additional 760 mm (30 in) from heavy cloud. Nagel (1956) attempted several different methods of measurement of the water content of sea fogs around Table Mountain but reported smaller catches. Where sea fogs drifting inland in summer are a nuisance, as in eastern Hokkaido, Japan, the filtering effect of pine trees has been put to good use (Hori 1953). Matsui (1956) measured a fog water content of 0.5 g/m^3 on the seaward side of coastal plantations of spruce and fir. This fog was effectively removed by passage through and over the trees. Air turbulence effects prevented any satisfactory quantitative description of the process. There is little doubt that the drip and stemflow benefit the roots of the trees themselves; Hursh and Pereira (1953) reported coastal forest in Kenya which was ecologically characteristic of 2000 mm rainfall areas, although the rainfall in the open was only 1143 mm. There is a substantial literature of observation, speculation and occasional measurement on this intriguing subject; the most recent review is by Kerfoot (1968).

In contrast to condensation mechanisms, the direct absorption of water vapour by plant foliage has been the subject of much technical controversy and experiment. The evidence is well summarised by Stone, Schachori and Stanley (1956), who show clearly that while water from mist or fog can be absorbed by pine-tree foliage and hence can assist in survival over dry periods, the water thus absorbed is not distributed to the roots, which die after some months of soil moisture exhaustion. Direct absorption of water from mist can therefore play a limited but occasionally useful role.

Interception of snowfall. Snow appears deceptively easy to measure and to manipulate as a water resource. In practice it proves very difficult indeed, as explained briefly in Chapter 3.

Because neither the initial snowfall nor the water content of the accumulated snowpack can yet be readily or accurately measured, the complex forest effects of interception, trapping, wind shelter, shade, re-evaporation and rates of melting are far from resolved, in spite of decades of observations in Russia and in North America.

Few snowstorms occur in the absence of wind, and since snowflakes fall at less than a metre per second, about one-tenth the speed of rainfall, they are usually drifting at an angle only a few degrees below the horizontal. When they arrive among the crowns of forest trees the pattern of eddies deflects many snowflakes into contact with the foliage, so that interception occurs more effectively than with rain.

Snow builds up rapidly on the foliage of perennials, especially on the needle-foliage of conifers; it may be blown off by wind, dislodged by falling snow from other branches, lost by direct mass release, washed off by rain, melted off or directly vaporised (Satterlund and Haupt 1970). The mechanical removal was found by Hoover and Leaf (1967) to be more important than the evaporation. They used time-lapse photography throughout the winter on pine and spruce trees of the Fraser Experimental Forest in Colorado.

Most experimental estimates of interception of snow by forest have been based on comparisons of snowpack in forested areas and in adjacent clearings. The differences may, however, be due in part to a redistribution of snow by patterns of air turbulence created by the clearings. Hoover and Leaf describe a pattern of forest felling designed to trap snowfall. Two-fifths of a 290-ha watershed were felled in a pattern of narrow strips. There was a 25% increase of streamflow, but very detailed snow surveys before and after the cutting treatment showed that there was no increase in total snow storage in the valley as a result of the tree felling. They observed that snow was blown off the trees and stored in the clearings. Measurements at over 1200 places showed differences in snowpack with an average of 100 mm more water equivalent in the clearings than in the trees. Costin et al. (1961) reported strong positive evidence of the accumulation and persistence of snow by tree cover in the Australian Alps.

Surveys of evidence on interception by Miller (1964) and by Satterlund and Eschner (1965) indicate that further progress will depend on detailed energy studies. Miller (1967) sums up the situation as follows: 'If the problem is properly conceived as a thermodynamic one in a

103

system of trees, snow masses and air streams, with their associated turbulent and radiative heat fluxes acting at several different scales, there is a good chance that success will attend a resolute attack on this area of hydrologic complexity.'

We must therefore leave this subject of snow-trapping as a quantitatively unsolved effect of forests on the behaviour of watersheds.

Effects of forests on snowmelt. Apart from the interception and trapping of snow, forests affect principally the rate of snowmelt in spring. A general observation is that snowpack under close stands of conifers melts some two weeks later than snow lying in adjacent open clearings. The protection of the soil from freezing and the high infiltration rates of forest litter permit more ready penetration of snowmelt water into the subsoil. In conditions in which arable or pasture soils are frozen and impermeable during snowmelt, the effects of forests, and even of broad windbreak plantations, can be important, as observed in the USSR (Chapter 4).

The immense range of geographical conditions and types of forest canopy prevents any generalisation and it is not surprising that apparently contradictory conclusions have been reached by different workers under the wide range of snow climates in the USSR (e.g. as surveyed by Molchanov 1960).

Winter streamflow from snowmelt caused by the flow of heat upwards from the slowly cooling soil can be of hydrological importance. Special provision of enclosed and heated weirs is needed when streamflow measurements are continued through frozen winters, but such equipment has shown steady flows from watersheds under a 2-m depth of snow, with air temperatures well below freezing (Federer 1965). The heat storage in the soil always takes part in the hydrological cycle of the watershed but only under snow is it readily demonstrated. The soil at the beginning of winter is cooled from the surface, so that a temperature gradient is created, with the soil warmer at increasing depths. Measurements with electrical resistance thermometers or with thermistors have shown soil temperature gradients of about 5 °C/m (3 °F/ft). Although the air temperature may be at -10 °C the base of the snowpack is usually near to 0 °C and the heat flow from the soil melts some snow. Where the soil is not frozen into a concrete ice formation this melt-water can percolate to the water table and maintain streamflow. Wisler and Brater (1959) surveyed the evidence and quote watershed yields of 0·25 mm (0·01 in) per day. Over large watersheds this can produce useful flows.

Effects of streambank vegetation. The water which percolates beyond the range of roots of hillside vegetation is safe from evaporation losses until the drainage flow approaches the river bed. Rapid drainage into a deeply incised channel may provide an efficient path, but where the stream is flanked by water tables lying near to the surface, often in alluvial deposits in which vegetation roots freely, an ecological association of plants well adapted to wet soils accumulates to exploit the water supply. The water use of highly adapted streambank vegetation is of major economic importance in south-western USA, where the tamarisk (*Tamarix pentandra*), known as the 'salt cedar', occupies thousands of hectares of the flood plains of major rivers. With roots either in the water table or freely supplied by the 'capillary fringe' films of water above it, these shrubs have full potential transpiration opportunity. Direct measurement of their effects is difficult, but the difference between the day and night streamflow rates in dry weather gives a rough estimate. Clearing of large areas of salt cedar in Arizona, with studies by the US Geological Survey of groundwater fluctuation, has indicated substantial savings of water.

Rowe (1963) reports dramatic increases in streamflow, of the order of 625 mm over the area cut, when riparian woodland was clear-felled in southern California. Dortignac (1967) gave an example of cutting costs at $4 to $5 per acre-ft (approx. 1200 m^3) of water saved, based on a study on the Pecos River. He argued that such water flows without further costs to reservoirs, treatment plants and transmission systems already developed and is thus likely to be less expensive than the development of new sources. Wildlife, fish habitat and scenery are all, however, jeopardised when streambanks are cleared, while soil protection is lost.

In South Africa forest plantations are not permitted within 20 m of streambanks, leaving a band of natural vegetation, but as Wicht comments, 'the dense phreatic vegetation, which usually develops on the streambanks, may account for vapour losses equal to those from plantations'. Rycroft (1955) and Nanni (1972) have demonstrated for South African mountain streams that clearing of streambank vegetation increased flow. Advection from dry grassland adds to the water loss.

Farmers developing irrigation schemes in East and Central Africa have urged the adoption of the South African rule for restriction of afforestation near to streambanks. That this cannot be applied usefully in the absence of a high water table is illustrated by Fig. 12, which summarises the relationship between trees and stream. Unless the trees and bushes are felled and regrowth is regularly cut or sprayed with herbicide, saving in water is small even on high water tables. Main-

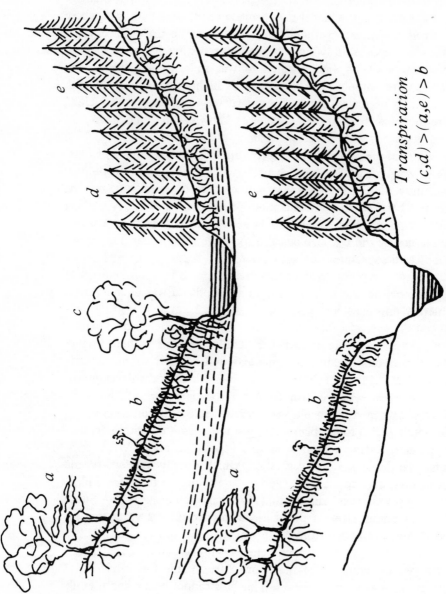

Transpiration
$(c, d) > (a, e) > b$

Fig. 12. The case for clearing streambank vegetation.

Planting of trees near streambanks is forbidden by law in parts of South Africa and similar legislation has been urged for other countries. The diagram illustrates that variations in rock structure can produce very different hydrological situations.

tenance of such strips as fire-breaks can sometimes be a solution, but in any case the extra water involves extra costs. The interesting question as to the allocations of such costs among downstream users of irrigation water was studied at the University of Arizona for the Salt River Project, in which a major cotton-growing area is irrigated from pine-forested watersheds. They found that, at a 6% rate of interest, it was economically feasible for cotton growers to pay up to $4·40 per annum for the first half-acre-ft (617 m³) of increased flow, with graded costs for larger quantities, in order to finance water-yield improvement measures and management in the forested watersheds (Rocky Mountain Forest and Range Experiment Station 1965).

Riverside strips of vegetation are also protected by law in the USSR, where margins 300 m wide are managed as protection forests. Rubcov and Dracikov (1970) argue that this width should be varied to include the whole of the river banks, the flood plains and any terraces flooded annually. This would involve a protection forest varying in width from 100 m to 1 km. They are concerned more with flood control than with water economy.

Control of stormflow. The general beneficial effects of forests on control of stormflow have been familiar from long experience, as described in earlier chapters. Watershed research has well established the power of forests to absorb heavy rainfall and transmit water to the soil by infiltration through the forest litter. A striking example, within the author's experience, was measured during the development of a tea estate from heavy rain forest in Kenya. Runoff from a 32-acre (13-ha) clearing developed on the contour for housing and administration was collected in a grassed channel and measured through an HL-type concrete flume of 110-cusec (3-m³/sec) capacity (Plate 11, Chapter 6). An equal area of forest was surrounded by a similar grassed channel.

In the heaviest storm experienced in this study, in which 3 inches fell in 30 minutes, at a peak intensity of 10 inches per hour (254 mm/hr), the flow rate from the administrative area reached 76 cusecs (2·15 m³/sec). The peak rate from the forest was less than 2 cusecs, which represented only half of the rain which fell directly into the grassed collecting ditch (Pereira et al. 1962).

In this same experiment, the peak flow rates from two valleys while under forest was 9 cusecs per square mile. When 600 acres out of the 1700 acres of one valley were cleared for tea planting, a heavy storm produced a peak flow rate of 373 cusecs per square mile.

Such extreme ability of forests to accept storms and to reduce spates must depend on a free movement of water through the profile and a large soil moisture storage capacity, as Hursh and Hoover (1941) demonstrated thirty years ago. Forest soils may be so porous that some water flow occurs before saturation (Hewlett and Hibbert 1963).

On shallow soils over impermeable rock the initial infiltration rate will still be high but rapid saturation of the soil will occur. Hursh (1943) pointed out these soil limitations in the forests of the Appalachian Mountains at Coweeta. A rainstorm of low intensity, totalling 8 inches (200 mm), falling on a wet catchment, caused stormflow of 30 cusecs per square mile from a forested valley with soils 6 ft deep or more, but 167 cusecs per square mile were measured from a forested valley with soils of only 2 ft depth. Thus the forest on shallow soil can delay the onset of stormflow, but when foliage, litter and soil are all saturated in a prolonged storm, the control ceases. Removal of the forest, however, merely accelerates the runoff.

Where either soil erosion or flood control are of major concern, simple and direct experiments can be of use to give early evidence of changes in streamflow response to rainstorms. Since these studies are concerned with the rate at which changes occur they need both rainstorm event and streamflow response to be recorded on time charts. For small experimental valleys of up to 200 ha (500 acres) a single recording raingauge and recording streamgauge can give useful results. Since the two records must be synchronised, and the clocks are the more expensive part of these devices, the author has found a Rhodesian pattern of instrument, which includes a raingauge on the roof of the stream-gauge recorder-house, operating a second pen to inscribe a rainfall trace on the streamgauge chart, to be a practical improvement. A second recording raingauge high in the catchment is needed to provide adequate cover for the purpose of stormflow analysis.

Ackerman (1955) demonstrated that for Pine Tree Branch Watershed, in the studies of the Tennessee Valley Authority, there was a fairly consistent relationship between both the size and intensity of rainstorms and the peaks of response in the streamflow.

Dagg and Pratt (1962) analysed the rainfall runoff data from the series of watershed studies in East Africa of which more details are given in the following chapter. From the charts of the recording raingauges and recording streamflow gauges they related rainstorm intensity and amount to the resulting stormflow peak in the streamflow.

The storm runoff was then predictable with good accuracy from the rainfall charts; e.g. over a year the storm runoff from two forested valleys, each of about 700 ha, was predicted within 6% and 7% of the amount calculated from the flow charts.

When the forest was felled and land use changed on one valley, the relationship changed sharply, while the valley remaining under forest showed no change (Fig. 13).

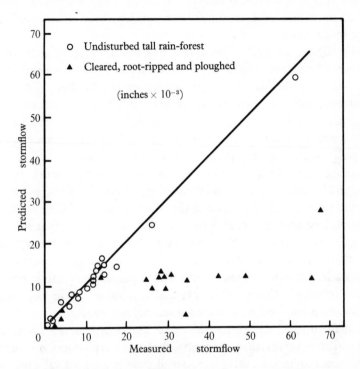

Fig. 13. **Prediction of stormflow from rainfall intensity and amount on Sambret Valley.**
Analysis of individual storms for amount × intensity over a year for two forested 700-ha valleys permitted subsequent stormflow to be predicted from rainfall charts within 7% of that shown by the streamflow charts. This relationship was altered sharply by clearing of the forest. The behaviour of the control valley (not shown) was unchanged.

Thus if major changes of land use are planned where flood control and erosion are important, simple pilot studies of this nature can yield a valuable measure of the hydrological consequences.

Integration of results by computer. It should by now be clear that there are many geological circumstances in which no water balance is possible. There are also many component problems of the effects of land use, such as snow storage or the effects of riparian vegetation, which can usefully be measured without resolving the hydrological equation of the watershed as a whole. For decisions on long-term policy, where major investment of natural resources and of funds are involved, a combined water balance and energy balance, both for present land use and for valleys under experimental changes, offers the most immediate basis for prediction, while local streamflow comparisons, over decades, offer the most secure basis.

The digital computer has made possible new techniques for the analysis of watershed data. Fashionably termed 'mathematical modelling', the watershed response to precipitation is described by a series of equations, the constants or parameters of which are adjusted until agreement is achieved between the measured and the predicted output. This is not to be confused with the building of physical models, which are well established in hydraulic engineering although used more for designing of bridges, harbours and coastal protection work than for watershed studies. A good example of modelling on a watershed basis was the scale model in concrete of the Mississippi River Valley, operated by the US Army Corps of Engineers at Vicksburg, in which the delaying effects of vegetation in the flood paths were simulated with folded wire mesh. The model was adjusted successfully to give valuable early warning of floods in the lower river and to permit model tests of flood-relief diversions. The concrete channel was some 100 m long and its construction and operation were substantial achievements.

A mathematical description in which the constants are adjusted by trial and error serves, more elegantly, the same purpose as the wire mesh in the concrete model; this technique already provides a valuable aid to flood control on many of the world's highly developed rivers. Where reservoir storage and power supplies are involved as with the chain of dams in the Tennessee Valley, efficient operation would be very difficult to achieve without the computer. Construction of such models is therefore an exercise in both hydrological engineering and computer technology; as with all studies in the fitting of equations to data, the critical tests are not only goodness of fit but also ability to predict, from new sets of data, results which can be verified in the field.

The ability of the computer to carry out exercises in which possible effects of land-use changes are introduced offers a new field of research. Many 'models' of watersheds are under development as research projects. Here the objective is to be able to characterise numerically the hydrological effects of different land-use patterns – crops, pastures, hardwood forests, conifer plantations, etc. – so that they can be superimposed on the routine model of existing rainfall and streamflow to give a prediction of the hydrological consequences of changes in land use. Reynolds and Leighton (1967) have made a well-argued plea for such integration, since planning of water resources in heavily populated industrial countries such as Britain urgently requires such prediction.

The many uncertainties of studies over whole valleys emphasise the pitfalls of the computer approach. These are increased when separate facets of the water balance are measured on small sample plots elsewhere and added into the study. Parcels of data may be fed into the computer with numerical accuracy but their assembly depends upon the quality of the equations on which the computer programmes are written. Computer 'models' now use only crude simulation of the watershed processes. As field research yields a better understanding of these processes and computing techniques develop, better models will be evolved. These in turn will need to be derived from, and tested by, more detailed and comprehensive sets of data from experimental watersheds.

The US Department of Agriculture Hydrographic Laboratory has for many years been assembling data on the percolation rates and water storage capacities of the soil types classified by the USDA Soil Survey, on experimental watersheds where rainfall and runoff are recorded (Holtan *et al.* 1968). By combining these with empirical rainfall runoff relationships established by the US Soil Conservation Service (Glymph and Holtan 1969), a research team has produced a mathematical model of the hydrology of a watershed (Holtan and Lopez 1971). This computer programme accepts data for rainfall and for evaporation from Class A pans, and also information characterising soils, vegetation and physical features of the watershed. The model carries out a soil-moisture budget, accounting for inputs, storage and losses. Sufficiently comprehensive sets of data on watersheds are not yet available to test this model, but the monolith lysimeters at Coshocton, described on p. 60, have provided an interesting approximation. England and Coates (1971) used data from a four-year rotation (grass, grass, maize, wheat, for four years 1944–7) as a calibration period and then applied the equations to the data for the next four years 1948–51. With rainfall and evaporation-pan data as input, the predictions of evapotranspiration and of percolation for the second period are shown in Fig. 14.

For reasons explained in Chapter 3, the lysimeters tend to drain rather more slowly than a field profile and hence to record a slightly high transpiration rate; the goodness of fit of the predicted and observed results represents a striking achievement in mathematical simulation of the hydrological processes.

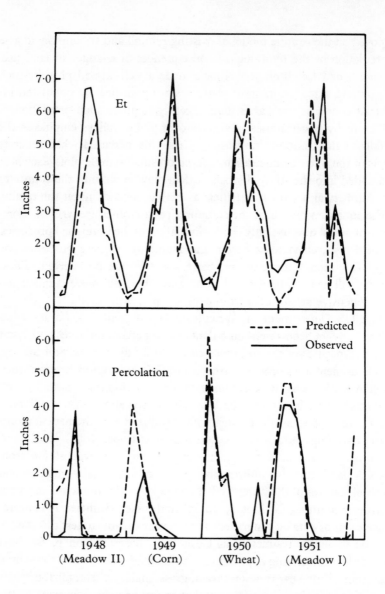

Fig. 14. **Prediction of evaporation and of percolation from a large lysimeter by a simulation programme on a digital computer.**

The USDA Hydrographic Laboratory has constructed a mathematical model of watershed processes which has shown encouraging success in the prediction of evaporation and of percolation throughout a four-year rotation on one of the large weighing lysimeters at Coshocton, Ohio. (England and Coates 1971; Holtan and Lopez 1971.)

In a proliferation of computer models by workers of varying experience and resources some of the results will be the outcome of false assumptions in writing the equations. It will become increasingly necessary to insist that they are tested against independent runs of good data, before being allowed to influence public policy. The simultaneous operation of such programmes with experimental land-use changes on watersheds equipped for research, offers the safest course. Substantial resources and determined interest by water-resource authorities are still necessary to resolve even the comparatively simple issues of afforestation in critical water-source areas.

Urgent needs for experimentation. Many of the world's most industrially advanced communities are in temperate climates, where geography is not very helpful to watershed experimentation, but where the need for quantitative forecasting of the effects of planned land-use changes on water resources is becoming increasingly urgent. While planning must needs be for decades ahead and pilot studies are a wise provision, as already urged in Chapter 3, large cities are already outgrowing the estimates of earlier planners and many major decisions on water resources have to be made before further long-term experimentation can bear fruit.

This urgency, and the recent development of equipment to intensify the studies, have led to a sudden proliferation of hydrological studies in Britain. Many of these are financed by the Natural Environment Research Council, which was itself established in 1965. A recent survey (NERC 1970) lists nearly fifty such research projects; many of them are joint enterprises between River Authorities and either major government organisations such as the Water Resources Board or university departments. Much of the field equipment is so newly designed that substantial effort is still needed to develop and test its reliability to the standard needed for routine use (McCulloch and Strangeways 1967). The Institute of Hydrology has devoted much effort to such tests and has established in Wales an intensively instrumented study of the water relationships of the adjacent headwaters of the rivers Severn, under forested plantation, and Wye, under pasture. Preliminary results for 1969 and 1970 indicated that the sheep pasture received 7% more rainfall and yielded 20% more runoff than did the forested valley (Rodda 1971). For early results the main effort must go into studies in which both annual water and energy budgets are assessed. The assumptions which have to be made in these assessments must in some cases await verification, but the multiplicity of information offers some safeguard against serious distortion of conclusions. Thus at a very late hour in the race to develop water resources at the pace demanded by the growth of modern industrial society, the staff and funds devoted to research are only beginning to match the urgency of the problem.

6

Watershed experiments in tropical forests

The foregoing chapters may have left the reader with more impressions of the difficulties and complexities of watershed studies than of their advantages. Some examples of the application of these studies in the developing countries may correct this impression.

The developing tropical countries are experiencing large-scale changes of land use as rapidly growing populations spread into previously uninhabited areas. The climates, soils and topography often present grave dangers of irreversible deterioration of soils and water supplies when land use is inappropriate or unskilful. These same geographical factors can also be highly favourable for watershed studies and offer valuable opportunities to obtain direct evidence in advance of major decisions on land-use policy.

An outdoor laboratory in high-altitude tropics

The advantages of a high-altitude tropical environment for watershed research may be summarised as follows:

Strongly contrasting wet and dry seasons annually, giving clearly defined soil moisture changes and streamflow recession curves.

High evaporation rates and low wind speeds.

Very fast growth of vegetation; several species of *Eucalyptus* grow over 50 ft high in four years, pine species reach 30 ft in seven years, tea bushes form a dense continuous canopy within five years, thus giving early establishment of experimental land-use changes.

Deep stone-free well-drained soils, of uniform physical structure and porosity. Soil moisture sampling, placing of electrical tensiometers and placing of access tubes for neutron moisture meters, are readily achieved.

Absence of snow and of frozen soils and hence freedom from the uncertainties which these produce.

Large areas under uniform vegetation in forest reserves giving uniform conditions of wind fetch and uniform administrative control.

Large areas of uniform geological formation, so that valleys of several km² can be found with the same major soil types.

It is also perhaps just to list some disadvantages in the form of difficult access by roads and tracks in the wet season (Fig. 15), sparse telephone communications and the need to learn to share forested catchments with fauna of formidable size and variety. Elephant, rhinoceros, buffalo, lion and leopard have all been encountered in the

Fig. 15. **Communications may be difficult in tropical mountain watersheds.**

Such difficulties are slight in comparison with the geographical advantages for watershed experimentation.

East African experiments described below but neither research staff nor resident observers carry arms; there have been no casualties, but various incidents of sharp excitement; there is always need for patience in such encounters, with prompt yielding of right of way.

A research opportunity. So far East Africa is unique in the development and the successful maintenance of such tropical experiments, using simple methods and employing field staff at a modest level of technical training. The studies are described here as examples of the type of measurements which could be usefully extended to many more developing tropical countries where empty lands are being settled. They have, however, some essential features.

Firstly they deal with immediate practical problems of land use for which governments are called upon to make decisions.

Secondly they are interdisciplinary and interdepartmental co-operative enterprises, calling into each small valley the specialists of a dozen or more professional and technical fields who otherwise often work separately.

Thirdly they have been the first priority for the energies of a small, full-time team of research workers who were aware of the results of watershed studies in other countries.

Fourthly although the experiments were, by modern standards, inadequately funded, they used simple equipment and the data were collected at little cost.

Watershed experiments in East Africa. Four studies have been set up and are continuing in Kenya, Uganda and Tanzania under what was initially the East African High Commission and is now the East African Community (a fifth study was later established in Zambia under different arrangements). All are continuing at the time of writing (1972). The central team was supplied by a regional laboratory, the East African Agriculture and Forestry Research Organisation (EAAFRO), set up in 1949 as part of the British Colonial Research Service to support the development of natural resources in the three East African countries.

Fifteen departments from three governments took part in the initial planning meeting in 1956 to which an active three years of preliminary reconnaissances were reported. Targets were agreed and responsibilities were accepted by the Water Development, Agricultural, Forestry, Geology and Administrative services of Kenya, Uganda and Tanganyika. Co-operation from a major international tea plantation company and from the East African Meteorological Department was also assured. EAAFRO supplied a full-time research team of three agricultural physicists, one biologist, three technicians and some ten junior staff.

Two studies of development of indigenous forest into plantations of economic importance were set up in Kenya for tea and for pine trees respectively. Two further studies concerned the improvement of eroded pastures in semi-arid country in Uganda (Chapter 7) and the effects of primitive cultivation on steep hillsides in Tanzania.

In spite of traumatic political changes involving the independence of all three countries, and the many consequent changes of staff, there has been continuity of policy and of research staff for this group of experiments. They have continued to receive active support from all three governments, each of which has reported them as a contribution to the International Hydrological Decade. The two Kenya forestry experiments have also been directly supported by the Overseas Development Administration through the Institute of Hydrology in Britain. Descriptions of these studies have been published in the research literature (Pereira *et al.* 1962; Pereira 1967; Blackie 1972). They are described here very briefly to illustrate the use of the methods.

Can productive tree plantations safely replace natural forest?
East Africa has many sad examples of the degeneration of perennial flows from naturally forested hills which have been unskilfully stripped of their forest cover. Torrential rain-season spates replace infiltration and storage, so that streams dwindle or vanish in the long dry season. The safe answer for the community is undoubtedly to protect the streamsource areas by Forest Reservations. These, however, cost money for fire protection and for policing, especially when the surrounding countryside is over-populated and over-stocked with cattle, sheep and goats. In spite of much forest research, commercial non-destructive harvesting of the natural forests has given yields too low to pay for this protection. When capital became available for investment in plantation enterprises, the unspoilt soils of the Forest Reserves offered the best prospects of success. There was then anxious debate as to whether it is necessary to lock up valuable land entirely to collect water, or whether modern scientific development on planned soil conservation principles can be undertaken without detriment to water supplies. The attractions, for developing countries, are the tax yield and employment opportunity of plantation enterprise in contrast to the cost of forest protection. In the more distant future, when storage dams are built, the question of total water yield will become more important. Until then, water quality

118

and regulation of flow are most needed. Plate 9 illustrates such a change in land use, in this case from forest to tea plantations.

Replacement of tall rain forest by tea plantations

The densely forested hills to the south of Lake Victoria receive a consistently heavy rainfall of from 1500 to 2500 mm. From this tall evergreen forest perennial streams supply large human and livestock populations in the lower and drier country of the lake littoral. Much of the forest has already been cleared and planted to tea, one of Kenya's

Plate 9. The South-West Mau Watershed Protection Forest.
The lower part of the dense rain forest has already been converted to tea estates. The land-use argument is about the remaining forest.

main export crops and an important source of employment (Plate 9). Capital investment, of the order of half a million pounds sterling per plantation unit, is very well protected by soil conservation work, but no prediction could be made with confidence about the effects on the seasonal pattern of the streamflow or on the total yield. With no storage reservoirs and no early prospects of building any, the critical test must be the effect on dry-season flow. Peak flows from the heaviest rains were also important since all road bridges were designed for the flows controlled by the forested headwaters.

The Kenya Geological Department described the south-west Mau Forest on the western slopes of Sitoten Mountain as underlain by a massive sheet-flow of hard phonolite lava (shown by a drill record to be 240 m thick) without any indication of fissures. The lava surface had weathered to deep porous stone-free soil.

Two parallel forested valleys were selected, after study from the air and from the ground. One of these, the 700-ha watershed of the Sambret stream, was excised from the Forest Reserve and leased to the Brooke Bond (East Africa) Company. The 540-ha completely forested Lagan Valley was studied as a control. The valleys selected were at a mean altitude of 2200 m with a slope of about 4%. A larger valley on which only streamflow was gauged was proposed for long-term comparisons.

After a preliminary year of comparison of the water balance of the two valleys, development of the Sambret began. Clearing and planting were planned and carried through to a timetable drawn up by the research group, the land-use change being carried out as rapidly as possible. One hundred and twenty hectares were cleared in the first year and planted in the second, and the whole area suitable for tea, comprising 350 ha or about half of the valley, was planted in four years with full development of roads, housing, factory and a small water-supply dam. Most of the remaining area lay within a subcatchment in the upper part of the valley, which was gauged separately and remained under natural forest.

Elaborate soil conservation precautions were taken. Plate 10 shows the early stages of the land-use change in which the careful contour planting, contour siting of factory and labour quarters and protection of the streambanks by indigenous forest are all clearly seen.

Beginning in 1958, this study has now been continued for fourteen years by EAAFRO and the Kenya Tea Company, with help from the Kenya Forestry and Hydrology Departments and from the Tea Research Institute; it has been sustained in recent years by essential specialist staff and equipment from the Institute of Hydrology of the UK.

Blackie (1972) presented to the IHD Symposium in New Zealand an analysis of the first eleven years of data. He reported that these have been assembled on punched cards, checked for consistency and transferred to computer tape as daily rainfall R, streamflow Q, Penman estimate of evaporation E_0, soil moisture storage changes in the root ranges ΔS and changes in water stored below the root range ΔG. Regular checks on ΔS have been maintained by sampling while ΔG has been estimated from the base-flow recession curves. The data have been computed on the water balance equation (or 'model')

120

Plate 10. A watershed study of a change in land use.

An experimental change on the 700-ha watershed which gives rise to a permanent stream, the Sambret. The subcatchment in the upper right-hand corner of the picture was separately gauged.

$$E_t = R - Q - \Delta S - \Delta G - L,$$

where L represents any possible net loss of groundwater other than by streamflow. The geological inference was that L should prove to be negligible. It cannot, of course, be measured directly, but three separate checks, described below, have confirmed that it is indeed either absent or too small to affect the outcome of the study.

The comparisons of E_t with E_0 have been used to compare the behaviour of the planted valley with the forested control and to relate their hydrological performance to other areas.

Results: water use. Blackie (1972) showed that over the first eleven years the mean annual water consumption was the same for both forms of land use. Approximately 1300 mm (51 in) per annum were evaporated both from tall rain forests and from the continuous canopy of tea and shade trees which replaced it. Both valleys gave an average E_t/E_0 ratio of 0·8.

The initial clearing gave an 11% reduction in water use. This was rather less than had been expected but only 30% of the valley was under bare soil at any one time. At Kericho the very favourable climate for tea growing is produced by rapid alternation between hot sunshine and rain (average of 6·5 hours of bright sunshine a day with 240 rain days per year). Van Bavel, Fritzchen and Reginato (1963) have shown by accurate experiments that while the soil surface is wet it behaves similarly to an open-water surface and the rate of evaporation is about the same. Blackie estimated the evaporation rate from bare soil to have been about $0·45E_0$ to account for the 11% reduction in water use by the whole valley.

Two exceptionally dry years, 1965–6 and 1966–7, had the unexpected effect of reducing the water use of the rain forest more than that of the tea plantations, so that over the eleven years, a period long enough to minimise the effects of annual storage changes, the mean water-yields of the two valleys were effectively equal (Fig. 16).

Checks for leakage. The checks for leakage begin with the energy balance. If the apparent water loss $R - Q$ is substantially greater than the potential evaporation from an open-water surface, as calculated from the Penman equation, then a leak is indicated. The possibilities of supply of advective energy from the air flow must be remembered including the special case of frequent wetting of the canopy so that only rates of loss over a whole valley of some 30% more than open water

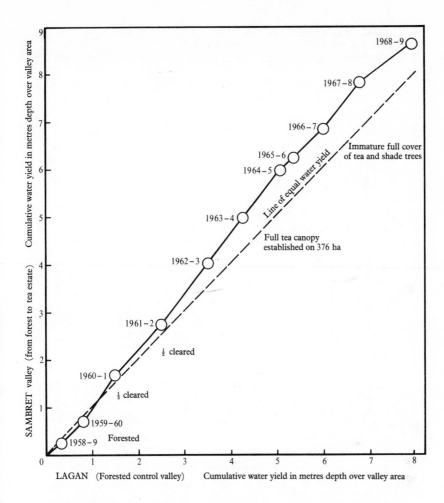

Fig. 16. Ten-year cumulative water yields from forest and from tea plantation.

Eight years after clearing began, the resulting increase in water yield totalled 1 m depth over the valley area, or about 2 m depth over the area converted.

would be interpreted as leakage. Two such experimental watersheds were thus detected as leaking in the East African experiments and were abandoned after two years, although the initial reconnaissances had not suggested such leakage. In both cases further exploration confirmed that leakage routes existed, bypassing the weirs. Although expensive this procedure is less costly than it would be to continue experiments for many years on the assumption that leakage could be neglected. Fig. 17 illustrates that for both the converted and control catchments the E_t/E_0 ratio has remained within the values expected for vegetation. In the

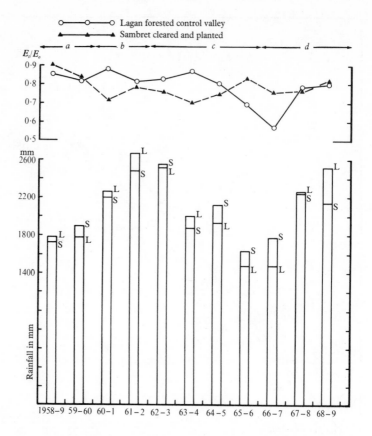

Fig. 17. **Rainfall and rate of water use, E_t/E_0 of forest and tea plantation.**

The forested controlled valley showed a steady rate of water use with an average E_t/E_0 ratio of 0·83, which was depressed in two drought years.

The treated valley was (a) under forest, (b) cleared and planted, (c) under a developing cover of tea bushes and shade trees, (d) under a full canopy of tea and shade.

forested control valley the steady values of about 0·83 are interrupted only by drought.

Secondly, with evergreen vegetation drawing on moisture stored in deep soils, the evapotranspiration can be expected to follow evaporative demand E_0, without serious departures from the annual E_t/E_0 ratio. A soil moisture storage balance, calculated on this assumption by the methods illustrated later in this chapter, has been compared with regular sampling checks of the first 3·2 m of stone-free soils. Such sampling cannot be replicated sufficiently to give an accurate agreement, which is

124

not necessary. If there is a leak, there will be a progressive departure of the forecasts from the samplings – no such departure was observed.

A third check has been possible by the use of an hydraulic weighing lysimeter, maintaining six tea-bushes as part of a level canopy. This has given, over the year 1968–9, an accurately determined E_t/E_0 ratio of 0·79 (Dagg 1970). This value Blackie compared with the value of 0·81 which he had obtained over the same period for the tea estate on Sambret, and that of 0·79 for both the forested subcatchment and the forested control valley.

There are thus good reasons to conclude that the geological indications are correct in this case and that the valleys of the Sambret and Lagan do not leak.

The conclusions from the first ten years of experiment are that the water use and the water yield, particularly the dry-season yield, have not been significantly affected by the completed land-use change (Fig. 17). These conclusions are, of course, valid only to the present stage of development, but further changes are expected to be slight.

Control of stormflow. In the absence of controlling reservoirs the effect of a land-use change on the regime of the streamflow is of critical importance. This was demonstrated in September 1960 when a very heavy storm, at the critical stage of maximum bare soil exposure, severely tested the soil-conservation design of the new plantations. This storm totalled 90 mm ($3\frac{1}{2}$ in), over 75 mm of which fell in one hour on a catchment already wet from ten days of rainfall; it had a peak intensity, registered by three autographic recorders, of a rate of fall of 250 mm per hour.

The flood control by the forest cover was most effective; the maximum stormflow rate reached only 0·6 m³/sec/km² (9 cusec/sq. mile) while the 600 acres of cleared and cultivated land delivered 27 m³/sec/km² (373 cusec/sq. mile). The well-constructed diversion ditches and drainage ways designed by the soil-conservation engineers withstood this assault and there was no serious damage or destructive soil loss. If such clearing were carried out on a larger scale, however, it would have a cumulative effect which would cause embarrassing peak flows in the lower reaches of the river.

Planting of tea bushes, of shade trees and of protective grass cover on roadside banks, all contributed to a rapid restoration of flood control. The effect of the development was to double the stormflow response to rainfall intensities below 12·5 mm/hr and to quadruple stormflow response to rainfall intensities exceeding 100 mm/hr. The fraction of

125

Plate 11. Concentrating storm runoff for measurement and discharging it into the stream.

Runoff from the 13 ha of the administrative area was gauged through an HL flume of 3 m^3/sec capacity. Water thus concentrated for measurement at the top of the streambank needed a heavily revetted cascade channel to avoid erosion.

the annual rainfall lost as stormflow remained very low indeed and after ten years of tea-estate development annual stormflow had fallen to the original forest value of only 1% of annual rainfall (Blackie 1972).

Runoff from the most intensively developed area of housing and factory, covering some 13 ha, was separately gauged through large concrete measuring flumes. The concentrated flow was returned to the river down a carefully protected cascade (Plate 11). This built-up area delivered 36% of the rain falling on it as immediate stormflow but these spates were fed into a small storage pond behind a weir created to supply the factory and this effectively controlled the peak flows and prevented disturbance of the stream.

Thus the critical stages in the development of land from protective forest to the cover of a tree crop have been shown to be possible without permanent deterioration in water resources either in quantity or in

126

behaviour. High capital input and the professional skill required to achieve this degree of hydrological control, almost equal to that exerted by the forest, is in sharp contrast to the unplanned invasion of forested catchment areas by peasant cultivators, which has destroyed most of East Africa's forests outside the boundaries of the forest reserves. It is important that the hydrology of this land-use change should be correctly interpreted. Mountain watersheds which are the source areas of important rivers need careful protection.

Natural forest, preserved against fire, felling and grazing, gives excellent protection. Tea estates, planned and developed with full soil conservation at a professional engineering level, have so far, over the first ten years, proved to be a hydrologically effective substitute.

A simplified approach: studies on plots of trees

Reynolds and Leyton (1967) have argued in favour of shorter-term studies on uniform plots of catchment area vegetation, to provide characteristic data which can be integrated by electronic computer to give estimates of watershed performance. This has attractions as providing interim data while the longer-term and more expensive whole-valley studies are in progress. Where climate and soils are highly favourable the short-term direct estimates are useful, as the following tropical example shows, but they are much more difficult under temperate climatic conditions.

On a uniform area of level land under a continuous canopy of trees, from which no overground runoff occurs, it is possible to find the water use directly from measurements of soil-moisture changes within the root-range. This assumes that withdrawal of water is by transpiration: it is therefore only valid where there is evidence for the absence of drainage losses. This evidence is readily obtained from simple and inexpensive electrical resistance tensiometers, called 'gypsum blocks' which indicate reliably when drainage ceases. Assumption of no drainage, without evidence, can lead to serious errors.

Soil sampling, at times dictated by the drainage conditions indicated by gypsum-block tensiometers, can be done with simple and relatively inexpensive apparatus, but requires much labour, both in the field and laboratory. These methods can now be replaced with advantage by neutron soil-moisture meters. The latter imply a higher standard of funds and of electronic servicing facilities than are yet available in many developing countries, and the simpler methods should not be prematurely discarded.

As a wise old watershed hydrologist in the USA once said to the writer, 'Many a field problem has been solved by the soil auger in less time than is needed to sort out the snags in a new gadget.'

From the rates of water use E_t between samplings compared with those of open-water evaporation E_0, a seasonal series of E_t/E_0 ratios can be established as for the Penman procedure. A continuous water budget can then be calculated from the records of rainfall and open-water evaporation, and this can be verified directly by further occasional soil sampling.

A fifteen-year test of a soil-moisture budget. By these methods the writer was able to demonstrate a continuous six-year water budget for a plantation of Arabica coffee trees in Kenya (Pereira 1957). A root study, prepared by a plant physiologist colleague, R. W. Rayner, showed the bulk of the roots to lie in the first 3 m of soil depth, and

Table 3. *Water budget for coffee trees (inches of water per annum).*

Budget calculated from open-water evaporation. Transpiration was limited by dry soil conditions in four years out of six; rainwater passed through the root range to recharge groundwater in only three years out of six. A continuous fifteen-year test has shown these ratios to give good prediction of water use.

		1951	1952	1953	1954	1955	1956
Rain	R	56	42	28	36	31	37
Open-water evaporation	E_0	62	61	63	62	60	54
Water use	E_t	28	37	33	29	34	34
Groundwater recharge	G	16	11	0	4	0	0

Monthly ratios E_t/E_0

J.	F.	M.	A.	M.	J.	J.	A.	S.	O.	N.	D.
0·5	0·5	0·5	0·8	0·8	0·7	0·7	0·6	0·5	0·5	0·7	0·6

sampling in this porous stone-free soil was to 3·2 m deep. When rain filled the soil profile beyond field capacity the surplus was assumed lost by drainage to recharge deeper storage. When the soil profile was depleted to wilting point transpiration was assumed to have effectively ceased. The study was undertaken as a basis for irrigation planning.

Table 3 summarises six years of results in which the sampling checks showed the calculation method to be fully adequate for irrigation purposes. The budget was continued for a further six years by Wallis (1963) as a background to irrigation experiments. He showed good agreements between the calculations based on Penman's E_0 and the sampling checks to 3·2 m depth, with 87% of the variation of the soil moisture predicted by the calculations. Since much of the difference is due to the real variability of soil moisture among tree roots this was a most encouraging result.

Blore (1966) further refined the irrigation control studies, and in doing so, carried on the original water budget of the unirrigated control treatments for a further three years, during which elaborate sampling checks were made. He took sixteen replicated sets of soil-moisture profile samples to 3·2 m every ten days for three years. Fig. 18 shows that the original (1957) basis of calculation still held good. Thus the method proposed by Penman and tested against short grass at Rothamsted gives practical results for coffee plantations near to the equator. The development of computer techniques has made the soil-moisture budget a practical tool both for control of irrigation and for study of water-sheds.

A soil-moisture budget for tea irrigation. In Malawi, where tea roots to 6 m deep in porous soils and flourishes in spite of a severe dry season, the methods developed for coffee in Kenya were applied by Laycock and Wood (1963); Laycock (1964) found by sampling in plots of irrigated tea that a ratio to the Penman evaporation estimate, E_0, of $E_t/E_0 = 0·85$ gave a good agreement with measured soil-moisture changes. When the tea was pruned and all leaf surface removed, transpiration ceased for forty days and averaged half rate for the next thirty days, after which the full rate of $0·85E_0$ was resumed.

Willatt (1971), in a very thorough deep-sampling programme combined with tension observations from gypsum blocks to a depth of 4 m, related rates of water use to the soil-moisture deficit below field capacity. Soil-moisture storage in the 4 m root range was approximately 550 mm of water available between the field capacity ($\frac{1}{3}$ atm) and wilting point (15 atm). Water use continued at about $E_t = 0·85E_0$ until a deficit of about 250 mm was reached. Transpiration was then progressively reduced as the deficit increased, falling to about $0·2E_0$. When heavy rain refilled the soil, water use was restored more slowly so that a plot of water use against soil-moisture deficit showed an 'hysteresis loop' (Fig. 19). This effect is one of crop response to climate rather than to

129

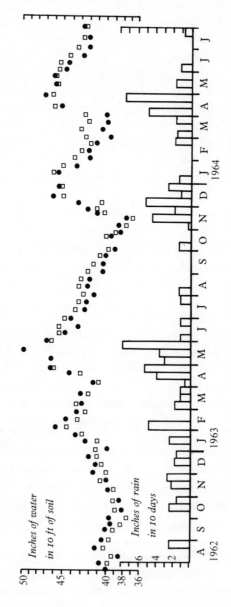

Fig. 18. **Soil-moisture prediction from evaporation rates and confirmation by soil sampling to 3·2 m depth.**

Soil moisture in the 3 m root range of Arabica coffee, predicted from evaporation rates by a cumulative budget method devised by Pereira (1957) (□), was confirmed in a thorough soil-sampling exercise by Blore (1966) (●). Soil capacity for available water and monthly values of the reduction factor f to give $E_t = fE_0$ are used.

soil properties. Willatt observed that calculation of a soil-moisture budget on this basis improved the prediction of the date of rewetting of the profile as confirmed by the gypsum blocks.

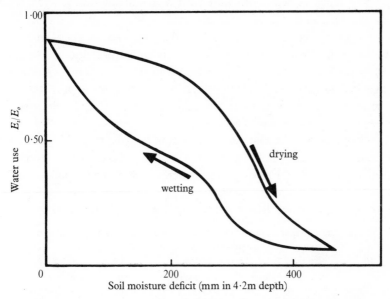

Fig. 19. **The effect of soil-moisture deficit on rate of water use by tea.**
Water use continued freely until approximately half of the available soil moisture stored within the root range had been used. The rate of transpiration was then reduced rapidly. When rain refilled the soil, the rate of water use was restored more slowly, giving an 'hysteresis loop'. (Willatt 1971.)

Bamboo forest or pine plantations in a mountain watershed

These simple methods were applied to a land-use problem in the forested mountain watersheds which supply Nairobi city. Trial plantings of pine and cypress had demonstrated this conveniently accessible area to be highly productive of softwood timber. The Forest Department proposed 40 km² of plantations, providing food and employment for 600 landless Kikuyu families, but were challenged on the water use of the softwoods as compared with the indigenous forest of bamboo thicket.

A reconnaissance study was therefore undertaken near the lower forest edge, where the warmer and drier conditions would give the maximum opportunity to demonstrate differences. Adjacent plantations on a level uniform site free from surface runoff were compared with the surrounding bamboo forest. These plantations, one mile inside the

131

forest, were presumed free from edge effects. The altitude was 2650 m (8600 ft); the long-term average annual rainfall was 1250 mm, with two dry seasons, and the annual open-water evaporation was approximately 1500 mm. In uniform plantations of Radiata pine 37 m (120 ft) high and Monterey cypress 16 m high, and in the bamboo thicket 13 m high, pits were dug and roots were washed with fire-fighting equipment to observe the depth and distribution. The main root development was within the first 3 m depth of porous stone-free volcanic soil, with negligible development beyond this depth. The horizons were sampled volumetrically and wilting-point and field-capacity measurements were made (Fig. 20).

Well-replicated profile sets of simple 'gypsum block' electrical resistance gauges for soil moisture tension were installed and resistances were measured on a robust portable meter with a hand-wound dynamo

Table 4. *Water use of trees by soil sampling.*

All three mature tree canopies used water at a similar rate, as shown by comparison with evaporation from open water.

	Height	Age	E_t/E_0
Natural bamboo thicket			
(*Arundinaria alpina*)	12 m	—	0·86
Radiata pine plantation	37 m	25 yrs	0·85
Monterey cypress			
(*C. macrocarpa*)	16 m	16 yrs	0·86

and an open logarithmic scale. From the drainage tension pattern thus observed it was possible greatly to reduce soil sampling occasions. Three four-man crews, in one day, were able to complete well-replicated patterns of soil-moisture samples to 3·2 m depth and to establish the soil moisture storage totals with acceptable accuracy.

The three forest types each drew water in most years from the full 3·2 m depth, while the resistance gauges showed drainage from the root range to have ceased; when the rains resumed, the deep soil took several weeks to refill to the point of through-drainage. Soil sampling under these conditions gave soil-moisture changes free from drainage losses. Water use by the trees was thus computed from the sum of rainfall and soil-moisture change.

Comparison of water use with open-water evaporation for the periods free of drainage beyond the 3·2 m depth gave the E_t/E_0 ratios shown in Table 4.

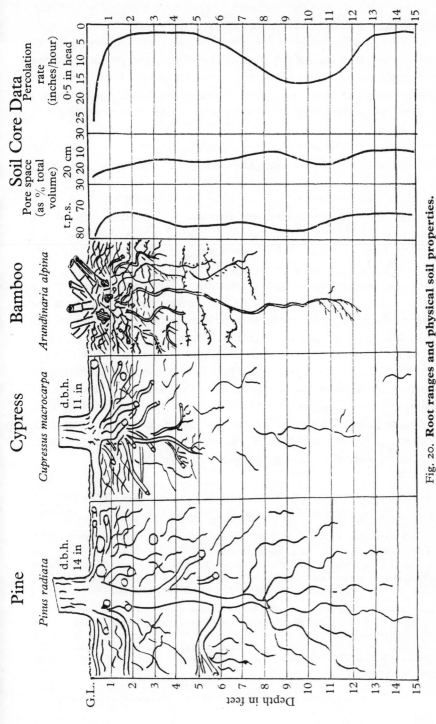

Fig. 20. **Root ranges and physical soil properties.**

When studying the water use of trees it is unwise to guess at root depths. Direct observation from root pits provides information on the soil profile and access for core sampling for volumetric measurements. Water jets should be used to reveal the root distribution, which is otherwise difficult to assess from the pit walls. (Diagram from Pereira and Hosegood 1962.)

The ratios for all three vegetation types were remarkably uniform. Little change would therefore be expected in the water use of the vegetation when bamboo forest is replaced by softwood plantations. However, in one long spell of dry weather the tensiometers indicated a more thorough drying of the soil profile under the tall Radiata pines, then thirty years old, than had occurred under the cypress and bamboo; soil sampling confirmed the difference as about 75 mm in the 3·2 m root range.

The measurements were continued while the cypress area was felled and replanted with Patula pine. This was done by the garden ('Shamba') system by which forest workers cultivate the soil and grow vegetable crops between the young pine trees. For the first three years the whole profile remained wet enough for frequent free drainage beyond the root range. Cultivation then ceased. Tall wood weeds (*Sambucus* spp.) rapidly filled the gaps in the pine canopy and created soil-moisture deficits slightly greater than those caused by the pines or cypresses.

From these simple studies it was thus possible to conclude that a short-term rotation of softwoods could be expected to use about the same amount of water as the undisturbed bamboo forest, with an expected increase of water yield during the first three years of clean weeding under vegetable crops, and an indication that an increase in the length of the forest rotation might lead to greater water use in drought years (Pereira and Hosegood 1962). The samplings were continued to compare age-groups of plantations and annual water balances were calculated. Table 5 shows that for this site at the warmer and drier extreme of the watershed protection forest there was effective recharge of deep groundwater in only four years out of eight.

Table 5. Water balance near the lower edge of a mountain watershed protection forest of bamboo thicket (Lat. 0°55'S, Long. 36°36'E, Alt. 2650 m).

A soil-sampling study one mile within the lower edge of the bamboo forest. The balance available for recharge is that calculated to have drained beyond 3·2 m depth. There was thus opportunity for effective recharge in only four years out of eight. The dangers of assuming groundwater recharge to be a constant proportion of the rainfall are evident.

	1953	1954	1955	1956	1957	1958	1959	1960
Rainfall	787	1295	940	1143	1372	1448	940	813
Water use ($0·85E_0$)	965	839	940	839	864	864	889	965
Balance available for recharge	−178	+456	0	+304	+508	+584	+51	−152

Watershed comparisons. The preliminary experiments, made at slight cost, had indicated in four to five years that, given short forest rotations, a major experiment could be undertaken without danger of creating a serious water shortage. The Forest Department therefore went ahead with planting and the earliest plantings were planned as an intensive watershed study.

The terrain, of steep valleys with deep porous soils derived from volcanic tuff and lava, was geologically unpromising of watertight catchments, but observations of the strong and even dry weather flows suggested that the valleys could be watertight. Core-sampling showed that at depths varying from 3 to 7 m, layers of dense plastic clay, sometimes less than a metre thick, effectively sealed these valleys. Weirs

Plate 12. Watershed protection by bamboo forest.
The measuring weir on the control watershed at Kimakia in Kenya. Data since 1957 have now been converted into computer code and assembled on tape. (Blackie 1972.)

were therefore constructed with concrete walls bedded firmly into the clay. Of the first three valleys thus equipped, two were subsequently shown by consistent water-balance and energy-balance calculations over ten years to have no indications of leakage. The third showed a water loss which could not be supported by the energy balance. Deep excavation eventually revealed an underground flow some 5 m below the weir foundations. This valley was discarded and serves as a salutary example of the need to check leakage.

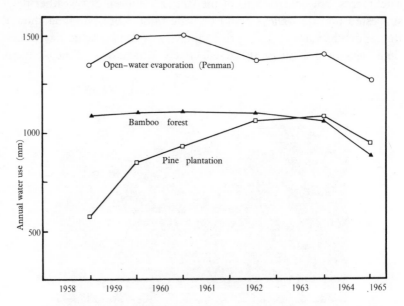

Fig. 21. **Water use of bamboo forest and pine plantation, compared with open-water evaporation.**
The clearing of bamboo forest and the establishment of young pine trees results in a saving of water for four years, after which the water use of the natural and planted forests reaches equality under the rapid growth conditions of the tropics. (Dagg and Blackie 1965.)

Since the best of the comparable pairs of valleys was already cleared for planting by the time these explorations had been completed, no preliminary comparisons were possible, but water-balance studies were relied upon to achieve an independent solution on each valley. Larger valleys with simpler gauging devices were recorded for purposes of long-term comparison.

The land-use change under study was from a protection forest of 12-m high bamboo thicket (Plate 12) to a forest plantation of Patula pine, maintained clean-weeded by the growing of peas, beans, squashes,

136

etc., between the trees until the pine canopy closed sufficiently to prevent further crop growth.

Measurements. As for the tea-plantation experiment already described, the variables measured included rainfall, from both autographic and standard raingauges; the data for the Penman estimate of evaporation were taken on site, and the calculation was corrected for altitude; soil-moisture tension was estimated by gypsum blocks and soil-moisture samples were taken at intervals on a volumetric basis to give an estimate of the depth of available water in the root range; a botanical survey was made on both valleys.

Results. The water use of the bamboo forest and of the developing pine plantation are shown in comparison with annual open-water evaporation in Fig. 21, for the first six years of the experiment, from data published by Dagg and Blackie (1965). The pines had then attained a height of 6 m and were using as much water as the bamboo forest. Thereafter the water use of the two canopies has remained approxi-

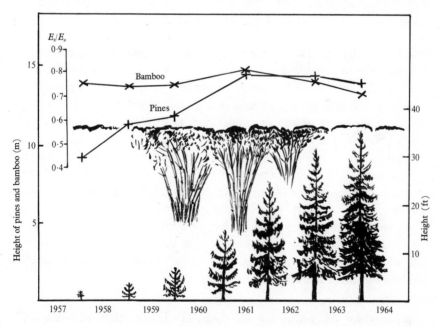

Fig. 22. **Rates of water use of bamboo forest and planted Patula pines.**
The rapid achievement of a closed canopy limits the period of water economy, but the high growth rate permits a short rotation in the felling and planting of softwoods, which has advantages in the management of watersheds.

mately equal, with an annual E_t/E_0 ratio between 0·7 and 0·8 according to weather distribution (Fig. 22).

Check for leakage. The gypsum-block tensiometers showed seasonal drying of the topsoil only, with ample reserves of available water in the root range of both pine and bamboo. Annual E_t/E_0 ratios were therefore used in constructing a soil-moisture budget on the principles already described on page 128. Since the water balances are cumulative, the predictions at successive soil sampling dates are not statistically independent, although they can be legitimately used to detect trends, such as would be introduced by any significant leakage out of the valley floor. However, in a series of separate dry seasons between each of which rainfall refills the soil profile, a statistically valid comparison can be made between predicted and measured maximum soil-moisture deficits. In ten successive cycles of soil drying and wetting occurring in seven years, the agreement between soil-moisture sampling and prediction was good enough to confirm that any leakage losses were negligible (technically a linear correlation of $r = 0·96$).

Up to an age of ten years and a height of over 15 m the water use of pine and bamboo remained equal, but the experiment must continue until the pines are large enough to be felled for pulp and thus to complete the rotation. The minimum economic rotation would be about fifteen years and it appears probable that this may be found to be hydrologically safe. It will be important not to forget, however, the indication from the preliminary experiments that an older stand of pines of a different species (Radiata) although using water at the same rate during wet seasons, had developed root ranges capable of extracting more water than the bamboo in a very dry season. The experiment should therefore continue in order to test whether there is a limit to the length of the rotation, beyond which the pine plantations might use excessive water in an exceptionally dry year. In the meantime the saving of the first four years of the rotation amounted to a total of 1550 mm over the planted area of the watershed, expressed as extra streamflow. For a city already provided with reservoir capacity for adequate storage such an addition to the water supply may be important.

Application of energy–budget analysis to the records of watersheds in the USA

Twenty-year records of the Soil Conservation Experiment Station at Coshocton, Ohio, whose massive lysimeters were referred to in Chapter

138

3, were studied by Mustonen and McGuinness (1968). They were able to compare the watershed records of rainfall and runoff with the readings of one of the Station's weighing lysimeters, a 60-ton column of undisturbed soil and rock under a permanent grass cover.

A small watershed of 18 ha, half under permanent woodland and half newly planted with pines, was compared with an adjacent valley of 123 ha under unchanged cover of pasture and scrub.

The frequency of wetting of soil and plant surfaces in the humid climate of Ohio was included in the study. Observations on the effect of hay-cutting in the lysimeters and direct experiments with plastic covers (Harrold *et al.* 1959; Patrick, Douglass and Hewlett 1965) have shown that in this region only about one-half of the total evaporation from soil and hardwood forest is due to transpiration. Mustonen and McGuinness made a multiple regression analysis of the twenty-year

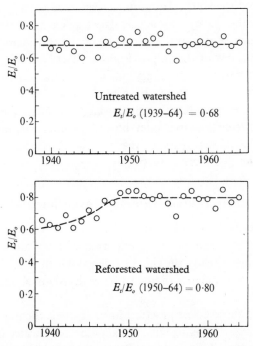

Fig. 23. **The water use of reforested and control valleys compared with evaporation E_0 from a theoretical open-water surface in the USA.**
The untreated watershed was under pasture and scrub. On the forested watershed one-third was under permanent woodland; two-thirds was under pines planted into poor grassland. After ten years of growth of the pines water use became stable at a rate about 18% greater than that of the control valley. (Mustonen and McGuinness 1968.)

records and found that the monthly evapotranspiration totals were significantly affected by the distribution of precipitation, but the effect on the annual totals was not significant.

They estimated open-water or 'lake' evaporation from the formula derived by the US Weather Bureau (Kohler, Nordensen and Fox 1955). This uses the same general energy-balance approach as the Penman equation, employing the same meteorological information of short-wave radiation, air temperature, humidity and wind speed. The geological situation of the two experimental valleys minimised the probability of leakage and evapotranspiration was estimated by the differences between precipitation and streamflow. The rainfall was estimated from nearby gauges, but only one gauge was within the experimental area. Corrections for annual differences in soil-moisture storage in the two valleys were made from a brief one-year soil-moisture sampling exercise related to lysimeter records and to well-level readings to give a rough index of soil moisture. The results, expressed as ratios of evapo-transpiration of the valleys to a calculated open-water evaporation (E_t/E_0), are summarised in Fig. 23. In spite of the considerable scatter, to be expected in view of the improvisations described, Fig. 23 shows a constant long-term evaporation relationship from the untreated water-shed. In contrast, the effects of the growth of a pine plantation are closely similar to those already demonstrated for Kenya in Fig. 22. In Ohio, however, tree growth is interrupted by winter, so that the pines planted in 1939 took ten years to reach a full transpiration rate, as compared with the five years observed in Kenya.

The future role of watershed experiments

The studies described in this chapter were set up to answer clearly defined questions of major practical importance. They are still in pro-gress, but they have already provided evidence which is of direct help to those charged with making difficult policy decisions about land and water resources.

In the course of these direct practical investigations the more detailed information about the ways in which vegetation uses water in stream-source areas is a bonus. This is quite different to the theoretical approach in which watersheds are elaborately instrumented in order to study their basic hydrology, and to advance man's knowledge of hydrological processes. The complexities of watershed studies already discussed render such studies increasingly difficult to support and they are appro-priate only to highly specialised and well-endowed laboratories. How-

140

ever, as forecasts of future water demands mount formidably in both tropical and temperate lands, far-sighted water-resource boards will have many clear-cut questions, which, foreseen a decade or more ahead, will permit practical studies of the hydrological effects of land-use changes to be undertaken in good time. For these, the simple techniques illustrated in this chapter offer some encouragement. Since both population growth and land-use changes are more rapid in tropical lands it is indeed fortunate that their geographical environments favour such measurements.

7

The effects of grazing animals on watersheds

Grassland as a land use

This chapter is not headed 'effects of grasslands', for comparison with the foregoing discussion of effects of forest, because the term 'grasslands' is too wide. The hydrological effects of grassland as a watershed cover depend directly on the land management, i.e. on the density and vigour of the grass cover and on the severity of the grazing, trampling and burning. The natural grasslands, variously described geographically as steppe, prairie or pampas, provide satisfactory watershed covers only while they are protected from overgrazing and from excessive burning. Irrespective of land use, flood problems can arise where spring snowmelt occurs over frozen subsoils, as frequently reported from the USSR, or where heavy rain falls on shallow soils overlying impervious strata. Water scarcity, as a result of long dry seasons and erratic annual rainfall, is a more general characteristic of natural grasslands.

Most intensively developed agricultural grasslands, in which grass is grown as a crop, are sited in areas formerly cleared from forest or woodland and under ecological conditions in which trees would again develop if management ceased. Vast areas of ranchland in semi-arid climates are also developed from natural dry woodland, scrub or savanna in which the trees and shrubs are held in check by rotational grass-burning or by more direct methods such as bush-breaking rollers, chain-slashing or spraying with chemicals.

The problems of grazing management and of hydrological stability are inseparable throughout this wide geographical span. Indeed in Mediterranean and sub-tropical climates, where subsistence agriculture and uncontrolled burning and grazing have largely eliminated natural forests, grazing management has a far greater influence than forest policy on the hydrological regimes of watersheds. Grazing practices are, however, deeply enmeshed in human behaviour patterns and are bounded

142

by land tenure traditions, so that improvements in land use are usually slow and difficult to secure. It is not, perhaps, surprising that few complete watershed studies of such improvements are reported.

The vegetation of grasslands has a deceptive apparent simplicity, since the water use depends not only on the climate but also on the depth of soil and the ability of the grass roots to exploit that depth. Root ranges encountered in watershed studies have varied, within the author's experience, from 5 cm to 5 m. Length of growing season, as with trees, sharply governs the annual water use.

Much confusion in the literature, and some extreme misconceptions in the transfer of technology from temperate to tropical climates, have arisen from over-simplifications by the civil engineers. With so little accurate watershed information and so many variables to consider, there has been a tendency to classify 'grassland' in hydrological calculations as a single form of land use with characteristic coefficients of infiltration and runoff. In temperate climates grass roots usually exploit only half a metre or less of soil depth and are not therefore liable to produce large seasonal soil-moisture deficits. In modern agriculture grass is treated as a major crop and is fertilised, grazed, mown and managed for maximum yield. This type of grassland affords a very high degree of soil protection, and has a characteristically high capacity for accepting rainfall: it thus provides a watershed cover from which flows either below or above the grass surface are suitable for reservoir storage. Water engineers basing their experience on temperate climates therefore often advise that for best water yield catchment areas should be planted as far as possible with 'short-rooted grass'. In less gentle climates, however, the soil conservation engineer has ample cause to recognise the wide variability of infiltration rates, of runoff and of erosion behaviour which can arise from grasslands. In semi-arid and tropical climates with higher rainfall intensities, less stable soils and longer dry seasons, the management requirements for watershed control under grassland are of critical importance.

Grasslands well supplied with water

Where water supplies maintain full transpiration rates for most of the growing season, grass can become a highly productive crop. Skilled management, with large dressings of fertiliser and intensive but carefully controlled grazing, can produce vigorous swards. Watershed studies in which poor grassland was improved to these conditions are reported from the USA, from Australia and from New Zealand.

143

At Watkinsville, Georgia, the planting of the deep-rooted Coastal Bermuda Grass (an improved strain of tropical 'Star Grass', *Cynodon dactylon*) with heavy dressings of fertilisers caused sharp reductions in both overland flow and in total water yield.

At Wagga Research Station in New South Wales two small watersheds, each of 7 ha, have been recorded for eighteen years by the Soil Conservation Service under heavy grazing regimes. While the control catchment area was given no other treatment, the intensively managed area was contour-furrowed, sown with annual ryegrass and subterranean clover, and given regular dressings of fertiliser. Both watersheds have been grazed with equal severity: over the eighteen years the improved pasture has maintained five times as many sheep as the control, with negligible stormflow or soil loss. The control valley has, over the same period, given thirty times the total water yield of the improved pasture (Knowles and Scurlow 1968). Similar experience was reported from Victoria, where Dunin and Downes (1962) applied fertiliser to a poor pasture and seeded it with annual rye grass and subterranean clover. This increased the stock-carrying capacity but halved the water yield. From New Zealand a combination of improved grasses and fertilisers on eight very small grassland watersheds (1·7 ha each) trebled grazing capacity but decreased runoff (Toebes, Scarf and Yates 1968). In such intensive grazing systems the continuous defoliation maintains an actively growing sward as compared with the seasonal seed production, ripening and drying of grasses which are allowed to grow to maturity (Yates 1971).

Effect of grazing on marshlands

After the Thames, one of the longest records of river flow in Britain is that of the Alwen and Brennig catchments in North Wales which contribute to the water supply of the city of Birkenhead. These are high moorland valleys, each of some 2500 ha, in a climate of water surplus. The vegetation is of coarse grasses and heather, with some exposed rock, some marsh areas of reeds, and some peat which was cut on a small scale for local fuel. The traditional land use was sheep grazing. Stream gauging began in 1922, after these two small catchment basins had been purchased for reservoir development. On the supposition that human and livestock populations in catchment areas would be sources of pollution, the hill-farming was discouraged. As tenancies became vacant they were closed and the buildings were demolished. Lewis (1957) studied the 33-year rainfall and streamflow records and concluded that the cessation of farming had led to the neglect of ditches draining roads and paths, and to the cessation of turf cutting for fuel. As a result, much of the land lay wetter throughout the year. Heather disappeared from large areas as the surface drainage deteriorated. The hydrological result has been a substantial decrease in water yield (Table 6).

Table 6. *Effect of cessation of sheep farming in wet moorland.*

Sheep farming was progressively reduced from 1922 onwards: drainage was neglected and water use increased.

Records of Brennig River	Rainfall		Streamflow		Water use	
	in	mm	*in*	mm	*in*	mm
1923–33	*53·94*	1370	*37·14*	943	*16·80*	427
1934–44	*50·25*	1277	*32·91*	836	*17·24*	441
1945–55	*52·04*	1321	*31·86*	809	*20·18*	512

Land use in grassland catchment basins of reservoirs under water-surplus conditions should clearly be managed for maximum grazing yield consistent with the preservation of a soil-protecting sward, since this is likely to increase runoff at the expense of transpiration.

Snow-trapping on cold rangelands

Mountain rangelands receiving substantial snowfall offer a further possibility of hydrological development. Trapping of snow by artificial barriers sited across the prevailing wind has been tested successfully over many years in alpine rangelands of the USA, e.g. by Lull and Orr (1950) in Utah and by Martinelli (1967) in Colorado. The increases of natural snow accumulation were equivalent to more than 3500 m^3 of water per 100 m of fence. Packer and Laycock (1969) estimate that this technique could be applied to some 40,000 km^2 of mountain grazing land in the USA, and that the trapped snow would increase late summer streamflow by an annual 1500 million m^3. Where such water can be delivered for use in developing valley irrigation the capital expenditure will be increasingly possible to justify.

Grazing in forested watersheds

A dense sward of well-managed grassland can withstand trampling of livestock and protect the soil surface. Where cattle are allowed to stray into woodland, with a tree canopy dense enough to suppress grass growth, they may often find sustenance by browsing, but the effect of their trampling on the unprotected soil surface causes rapid disturbance of the hydrological regime.

Gauging of streamflow for seven years at the Coweeta Hydrological Laboratory, in an ecologically stable area of hardwood forest, was followed by grazing of only seven head of cattle in 145 acres of forest. For a few seasons there was no detectable effect, but after seven grazing seasons surface runoff channels were established and erosion began on a serious scale. Maximum flows increased by four times (Johnson 1952). In another experimental basin in which the trees had been cleared and the land had become rough pasture, the streamflow was calibrated against a forested control for seven years. The maximum flow was below 1 $m^3/sec/km^2$. The valley was then opened to the unskilful grazing characteristic of local land use. Within four years flood peaks had increased by ten times and a maximum flow of 20 $m^3/sec/km^2$ was recorded.

The grazing of cattle in forested water-source areas must therefore be confined to the controlled use of the grass-covered glades; where the ground cover is forest litter the livestock must be kept out of the trees.

The need for such discipline in the grazing of streamsource areas was demonstrated early in the history of the USA, as a result of the very rapid settlement of the West. In Utah, uncontrolled grazing of mountain watersheds led to disastrous torrent flows of soil and rocks on to the townships below. Croft and Bailey (1964) have reviewed the evidence in a well-illustrated survey. They describe large-scale restoration treatments by bulldozing of contour trenches to arrest runoff and by reseeding with vigorous grasses to restore infiltration rates. These treatments gave highly successful control of summer floods. The dry-season streamflow from the restored grass cover was greater than from comparable areas of woodlands (Bailey, Craddock and Croft 1947; Bailey and Copeland 1960).

A planned land use of forested slopes and grazed valley floors can result in economically and hydrologically productive watersheds. This is well demonstrated in the closely controlled mountain regions of Germany, Switzerland and Austria (Francois 1953). The Forest Service of the USA licences very large areas of grazing in water-source protection reserves. The first protective legislation was passed in 1891. Today about one-quarter of the Mountain Region of the USA, which covers most of the twelve States from Mexico to Canada, is National Forest land, and together with State Forests totals some 55 million hectares (138 million acres). Of this great area of protected watersheds 44% is grazed by livestock under a fully controlled system of paid licences (Cliff 1958). Over one million cattle and 2·7 million sheep enter the forests annually on permits for summer grazing. Rotational grazing for the protection of the soil and the preservation of palatable grass species is assisted by range fencing (30,000 miles), by stock-trails (2,500 miles) and by the provision of water points (19,000). This capital development is provided mainly by the Federal and State authorities with contributions from local 'grazing associations' of permit holders. These local associations work out grazing rotations with the range-management specialists of the Forest Service. They also help in the active improvement of range grazing by the clearing of unwanted juniper scrub and by reseeding with palatable grasses.

In streamsource areas where surface streams and shallow groundwater provide readily available drinking points such paddocking schemes are easily arranged: control of stocking rates in the interest of hydrology is more difficult. Growth of cattle forage varies greatly from year to year and a firm over-riding control is needed to remove excess stock in years of scarcity of grass. The importance of exercising such control through local associations of graziers is that experience demonstrates the profitability of good pasture management, so that the interests of the graziers and of the hydrologists are eventually seen to coincide. This policy is of great importance to developing countries where a combination of firm control and voluntary participation is equally essential among pastoralists with different traditions of grazing practices.

Grazing by forest wildlife

Foresters and game rangers are perennially in dispute about the destruction of young trees and the critical damage to the bark of mature trees by the larger wild animals such as elk and moose in North America and elephant in Africa. Overpopulation with any species, from man to mouse, can have destructive effects, but it is quite surprising how little hydrological damage arises from wild animal species in forests when not constrained by human intervention. The most dense concentration of wildlife in tall forests known to the writer is that around 'Treetops', the renowned game-viewing centre on the slopes of Mount Kenya. The concentration of elephant is here artificial in that a traditional route between the forests of Mount Kenya and the Aberdare mountains is interrupted by a game barrier (a ditch 3 m deep, screened by a cover of brushwood) to protect the maize crops of an African settlement from elephant damage. Muddy water in a small stream issuing from the forest at this point was attributed, by a local forest officer, to the effect of elephant, rhino, buffalo and other species. The local Game Warden insisted that the larger animals did not drink at this stream. The writer accompanied the two of them in a hair-raising walk up the stream using a shilling clipped by a bent wire to a metre stick as a simple turbidimeter, since both protagonists agreed that the depth at which the coin vanished was a practical measure of the clarity of the water. The Warden led us to the water-holes; these were marshy pools below salt licks, embarrassingly well populated. The ground around them was trampled into deep mire but the pools had no direct drainage into the stream. Following the stream we eventually found the point of pollution to be a deep gully developed from an old contractor's logging trail angled sharply downhill, which had captured a tributary flow and was draining a continuous stream of muddy water into the small creek. This damage, persisting from the extraction of a few valuable timber trees among a dense forest, emphasises the need for close management of forests for watershed protection and the necessity to investigate causes rather than to dispute about effects. The damage caused by logging in forests of the USA was studied by Packer (1967) in 720 operations. He proposed practical remedies.

Grasslands having long dry seasons

Contrasting with dense forest there are far greater areas of lower rainfall in which, without interference from man, a natural balance prevails between trees and grasses. The vegetation of high-altitude tropical and sub-tropical Africa is predominantly of this character. Many ecologists believe that this 'parkland' association of trees and grasses is a 'fire climax' maintained over centuries by grass fires started by lightning strikes from electric storms. The frequency of such fires was increased by early human settlement, through deliberate burning in support of either hunting or grazing activities.

Even under modern conditions of rangeland management, deliberate grass-burning is the main method used to check regrowth of bush; ring-barking and use of arboricides offer less destructive but more expensive alternatives (Ward and Clegghorn 1964).

The climates in which such associations of trees and grasses develop have characteristically long dry seasons with shorter and somewhat erratic rain seasons. This variability causes universally difficult problems of range management. Years of low or poorly distributed rainfall give poor grass growth; shorter wet seasons automatically increase the duration of fire hazard in the longer dry seasons, destroying the reserves of dry forage.

In open woodland-savanna areas not yet over-run by human population development, large wildlife populations of animals graze and browse a wide variety of grasses, herbs, shrubs and trees. Active management control of the numbers of wild herbivores in protected areas is essential to avoid accelerated overland waterflow and soil erosion. The measures to achieve this have already been outlined and references quoted in Chapter 2, where the advantages of protecting critical streamsource areas by National Parks were emphasised.

Ranching enterprises with domestic livestock are profitable in the long term only when managed to avoid critical damage to the grass cover. When man replaces the wildlife by dense concentrations of herded livestock of only two or three species, the liability to damage in drought is greatly increased. As science has brought improved health protection to man and beast, population densities have increased, with hazard in many developing countries to hydrological regimes of the range watersheds. Can the water relationships be maintained, under the increasing populations of men and their livestock, to secure the infiltration and subsurface flow, i.e. can intensive grazing be maintained without accelerating both overland flow and soil erosion?

The problem is acute in India, where Whyte (1958) has described the damaging uncontrolled impact of one-quarter of the world's cattle on forests and grasslands, but the literature does not appear to contain critical data of the effects on the watershed hydrology. The experience gained in a watershed experiment on severely overgrazed land in East Africa is worth quoting, although the experiment is still in progress.

Watershed experiments in range improvement

A study in East Africa. This study of restoration of overgrazed savanna country is in the Karamoja district of Northern Uganda, at the top of the head-water catchments of the Nile near to their boundary with the watershed of Lake Rudolf. The annual rainfall of 500–750 mm, although far less than the open-water evaporation of some 1800 mm, is adequate to maintain open woodland with a rich flora of grasses. Well-developed trees of many species of *Commiphora*, *Acacia*, *Terminalia*, *Balanites*, and *Tamarindus* dominate a rich flora. This country can be productive as ranchland, but persistent misuse by overgrazing has reduced many hundreds of square miles to the condition illustrated by Plate 13a. Ecologically, the vegetation has become thornscrub with desert grasses. Runoff from the bare and trampled soil is so great that men and their livestock depend precariously on seasonally declining waterholes dug in dry sandy watercourses. Hydrologically, the results are embar-

Plate 13*a*. Results of persistent overgrazing in the Uganda headwaters of the Nile.

This woodland savanna under an annual rainfall of 500 to 750 mm can provide productive grazing, but persistent misuse has reduced very large areas to the condition illustrated.

Recovery is illustrated in Plate 13*b*.

Plate 13*b*. Recovery of overgrazed watershed.

In spite of severe loss of topsoil by sheet erosion, bush-clearing and resting resulted in a vigorous grass cover, in which fifty-nine different species were identified.

150

rassing to the road authorities, since bridges over the drainage lines are repeatedly damaged by torrent flows.

The primitive pastoral tribes have not begun to understand their own predicament. They attribute their troubles to 'poor rains' while their increasing numbers of livestock relentlessly reduce their habitat to desert. Rainfall records for thirty years at Moroto give no indication of decline. Construction of extra watering points by building of earth dams and drilling of boreholes was attempted by the administration in order to spread the grazing over wide areas, but cattle populations grew rapidly around the new water sources and merely extended the area of destruction. The development of further water points was therefore stopped pending the achievement of grazing control.

A watershed study of the restoration of a typical area was begun in 1956 at Atumatak near Moroto, as one of the series of co-operative experiments between the East African Agriculture and Forestry Research Organisation and several government departments, as described in Chapter 6. Full descriptions and initial results have been published (Pereira *et al.* 1962).

The topography is gentle and the land use is typical of a very large area. A flat shallow basin of 8 km² (2000 acres) is divided by a rock ridge into two similar watersheds. The whole area has been continuously overgrazed and eroded for many years. Extensive sample quadrat counting gave the average exposure of bare soil as 40% of the watershed area for nine months of each year. There is no permanent streamflow and a typical observation of response to a rainstorm of 20 mm is a brief violent spate ceasing abruptly after 30 minutes. Watershed research under such circumstances must be concerned with the amount, distribution and intensity of the rainfall, the spate or stormflow and the depth of penetration of the rain into the soil of the grazing areas. Standing-wave flumes of 30 m³/sec capacity were designed and built by the Uganda Water Development Department; these were fitted with heavy steel pipes set in concrete to provide guide rails by which current meters were raised and lowered by a winch during stormflow in order to confirm the theoretical calibrations of the designs (Plate 14). Rain gauging was intensive, with one daily-read gauge per 40 ha and one recording rain-gauge per 160 ha. Depth of penetration of rainfall and the duration of available water in the soil profile were recorded by twenty sets of electrical tensiometers (gypsum blocks). Evaporation was estimated both from pans and from data for the Penman equation, recorded on site.

For a calibration period of four years both valleys remained under continuous heavy grazing while the relation of rainstorms to spateflow

151

Plate 14. Standing-wave flumes on grazed watershed.
For soil-laden torrent flows these three-stage flumes provided measurements up to 30 m³/sec. Low flows were gauged through the Parshall Flume in the centre.

was studied. On both valleys storm runoff was found to be closely predictable and to average 40% of all precipitation.* The first two or three storms at the beginning of each rain season gave lower runoff as the sand beds in the torrent channels were recharged. Penetration of rainfall into the trampled soil was slight; on many of the bare soil sites the gypsum blocks showed no wetting deeper than half a metre. The gypsum blocks were sited near to raingauges and soil cores were taken nearby. Calculation of water stored, based on the depth wetted to field capacity, indicated that at least one-third of the rainfall had been lost by runoff. This agrees well with stormflow measurements.

While excessive grazing continued on one watershed the adjacent valley was treated by bush clearing and by controlled grazing protected

* For those technically interested the erratic distribution of rainstorms was best measured by dividing the watershed area A covered by the twenty gauges, into strata of area a with two gauges per stratum.

If R_a is the rainfall received by this stratum, the runoff depth over the watershed is given by $y = (0.4/A)\sum a\,(R_a - 0.35)$. The surface deficit of 0.35 inches often developed in a single day. For the thirty-nine significant storms occurring in 1960 the linear regression of measured runoff on estimates of runoff by the above equation accounted for 93% of the variance (Pratt 1962).

by fencing. The bush was chain-cleared by heavy tractor and bulldozed to the perimeter with a root rake (a series of tines bolted to the edge of the bulldozer blade). For the first season cattle were excluded. Although much of the topsoil had been lost by erosion and a mantle of stones littered the surface of bare subsoil, no reseeding was found to be necessary. A rich flora of grasses developed over the battered valley (Plate 13*b*). The botanist member of the team undertook thirty one-acre quadrat counts and collected fifty-nine species of grasses of which thirty-seven were perennials (Kerfoot 1962). These had been protected by dense patches of thornbush and by clumps of *Sansevieria* spikes which had defeated even the hungry goats. Penetration of rainfall into the soil was increased from 0·5 m to 1·25 m depth, while peak flows were reduced. After observing grazing effects on the restored watershed, experienced range specialists now estimate the carrying capacity of the cleared and rested land at one beast per 6 acres per annum (one per 2·4 ha). This is nearly double the grazing yield of the two watersheds before treatment, when daily counts of cattle, donkeys, sheep and goats gave a yearly grazing average equivalent to stocking with adult cattle at 11 acres per beast. That application of simple pasture management techniques will simultaneously improve both water infiltration and grazing has thus already been demonstrated. The mechanism of the improvement is illustrated by a study under similar conditions of the rainfall penetration into the soil below single clumps of grass (Glover, Glover and Gwynne 1962).

Unfortunately a period of tribal disturbances and cattle raids then followed in which the experimental cattle were repeatedly lost and the grazing regimes were therefore erratic. The streamflow measurements and meteorological records were maintained; control of bush regrowth by slashing was continued. A recent re-establishment of controlled grazing on this experiment, stimulated by the interest aroused by the International Hydrological Decade, has permitted the initiation of a new set of stormflow records under known grazing conditions from which a quantitative hydrological measure of the improvement will be possible.

Regeneration of tropical grasslands. The wide botanical range of grass species characteristic of the semi-arid tropics affords a degree of resilience which can permit recovery from desolate conditions. Reseeding is a characteristic technique in restoring eroded grasslands in higher latitudes (USA, Australia, New Zealand) and is often advocated by visitors to Africa. In a direct comparison of methods of restoring misused and eroded land in Kenya (Pereira and Beckley 1952) it was found that deep ripping led to drying out of an excessive depth of soil and retarded recovery, but that contour ridging was beneficial. Reseeding after exhaustive arable cropping was most successful when local indigenous Buffel grass seed (*Cenchrus ciliaris*) was used. Further study of the productivity of the natural grasses, when protected from grazing and allowed to regenerate, showed that full recovery took two years longer, but the grazing capacity was then as good as or better than that obtained from ploughing and reseeding (Pereira, Hosegood and Thomas 1961).

Evidence from Australia. Although the arguments for tillage and reseeding are agricultural, the outcome is of urgent importance for the control of runoff and erosion from grazed watersheds. This is demonstrated convincingly in Australia by the Soil Conservation Authority of Victoria. When the State Government of Victoria voted funds in 1960 to build the Eppalock water supply dam to store 300 million m^3 on the Campasbe River, a study by the Soil Conservation Authority showed an area of 500 km^2 to be severely eroded by sheep grazing. For the first time in Australia's history, funds were voted for soil conservation in the watershed as part of the dam-building project (Downes 1961). A ten-year programme has been carried out jointly by the Soil Conservation Authority (1966) and the farmers. Fortunately the eighty-year records of riverflow maintained in Victoria afford an opportunity to measure the effects of this land-use change. With the co-operation of the Soil Conservation Authority the writer was able to study both the watershed and the records, in the course of the IHD exercise referred to in Chapter 4.

The history of early gold mining in this area led to a sharp contrast in land use between the two parallel and adjacent valleys which drain into the Eppalock Dam. The Loddon River valley has been lightly settled and has not been excessively grazed; much of the watershed is still covered by the indigenous eucalypt woodland. It therefore forms an excellent 'control' for the Campasbe River valley which has had a less fortunate history. The discovery of gold in the sedimentary soils at the lower end of the valley led to a 'gold rush' and the concentration of a heavy population from 1850 to 1860. The small village of Heathcote grew to a town of 35,000. Firewood cutting, both for the town and for the ore-stamping machinery, cleared large areas of eucalypt woodland in the Campasbe watershed. When the gold deposits were exhausted many of the miners earned their living by cutting firewood for sale to the growing metropolis of Melbourne, which further depleted the woodland cover. Heathcote's population fell slowly to 1300 and the land use changed to sheep ranging on the cleared areas. There was no effective grazing control, for sheep give saleable wool even when underfed, so that stock numbers were limited mainly by survival of drought. The adverse economic climate of the 1930's was accentuated by seven unusually dry years out of ten from 1936 to 1945, so that the graziers had little to spare for remedial measures and farm improvement. The denudation was completed by a plague of rabbits. Only one or two pioneer homesteads succeeded in maintaining stability through careful husbandry. Attempts at restoration of the denuded land by seeding and by the application of fertilisers ran into unexpected technical difficulties, for although the need for phosphatic fertilisers was demonstrated by soil analyses, reseeding and fertilising with phosphates failed repeatedly to re-establish grass cover on bare eroded soil. Research into soil fertility pinpointed the critical factor in 1950 as a deficiency of molybdenum and the necessary small quantities of molybdate were added to the superphosphate fertiliser. Biological research also led to the elimination of the rabbit plague by the crude but effective dissemination of the myxomatosis disease. The selection and development of the invaluable Subterranean Clover (*Trifolium subterraneum*) and a hardy and productive pasture grass (*Phalaris tuberosa*) provided a successful mixture for the restoration of pasture land.

When the Victoria State Government decided in 1960 to harness the water resources of the Campasbe River for power generation, the technical difficulties of restoring the eroded catchment had been solved but the practical problems were severe. Five hundred km², with a rainfall of 650 mm, had been grazed to the stage of severe sheet erosion with less than 50% of ground cover. The sedimentary soils were unstable when exposed and were prone to a particularly vicious form of soil erosion by tunnelling in which the tunnels subsequently collapsed to give steep gulleys. Active co-operation with the landowners was essential. The farmers contributed seed and fertiliser while the government grant paid for chisel ripping on the contour to provide a seed bed, the creation of diversion ditches and concrete drop structures to lead excess water safely to the river and the stabilisation of river banks and gulleys with gabions of rock and wire netting. Both tree planting and extension advice to farmers on management of grazing were included in the restoration programme (Plate 15).

The effect of the cumulative deterioration of the river flow regime under severe large-scale overgrazing is clear from the records maintained at Elsmore, a short distance downstream of the Eppalock Dam, since 1885. For the first twenty years, 1885–1905, before severe grazing

155

began, the annual peak flows remained below 1 m³/sec for each 10 km² of catchment area (9 cusecs/sq. mile), although annual rainfall varied from 25% below to 40% above the long-term mean.

After 1905, as grazing pressure increased, flood peaks increased up to a maximum of five times the previous record. Nineteen times in the next

Plate 15. Repair of gulley erosion.

(*a*) Two concrete drop structures to convey water to the bottom of a gulley without cutting back into the soil. Over 300 have been built in the Eppalock Catchment.

fifty-five years the flood peaks of the first twenty years were substantially exceeded. On the neighbouring Loddon catchment, of 4000 km², the flood peaks exceeded the pre-1905 levels only five times in the same period of fifty-five years. Construction of the dam ended the flow comparisons but it also made possible a direct estimate of the transportation of sediment. Previous to 1960 sediment flow estimates from river sampling by the State Rivers and Water Supplies Commission agreed well with an independent estimate by the Soil Conservation Authority based on measurements of the growth of gulleys seen in successive air photographs over a ten-year interval. The two estimates agreed at about 10,000 m³/year of transported soil and rock (for comparison with the USA data on the assumption that the sediment was derived mainly

(b) Tree planting to stabilise banks of gulleys. 108,000 trees have been planted in 51 miles of gulleys in the Eppalock Catchment. (Photographs by Soil Conservation Authority of Victoria.)

from the badly eroded land, the rate of erosion would have been of the order of 5000 tons/sq. mile/year). The very dry year of 1967 permitted a survey of five years of sediment accumulation in the Eppalock Reservoir. This represented only one-quarter of the previous rate, denoting a very rapid improvement under the concentrated programme of soil conservation work and pasture improvement.

Water use by improved rangeland. Two opposing hydrological factors can be accepted as established by evidence: firstly, that control of grazing to preserve a grass cover and to prevent excessive soil exposure and trampling is an essential precaution against accelerated soil erosion and flood flow; secondly, that improving of the density and productivity of grassland decreases total water yield. As with forest cover the protection of soil, preservation of water quality and moderation of flow patterns are obtained only at a substantial price in extra water use by the vegetation.

In the case of deep-rooted perennial grasses this price can be very high indeed in climates with long dry seasons where total annual evaporation exceeds annual rainfall. At Muguga in Kenya, analysis of the water balance under well-managed 'star grass' (Bermuda grass), *Cynodon dactylon*, in grazing experiments where rainfall, runoff and soil moisture changes to 3 m depth were measured for six years, showed how serious the hydrological results could be. Even on 10% slopes these pastures gave negligible runoff and used all of the 990 mm (39 in) of annual rainfall. Star grass newly planted on a deep soil at field capacity was shown by soil-moisture tension indicators (gypsum blocks) and by soil sampling to be extracting water down to 3 m depth within the first year and to maintain continued subsoil moisture deficits which effectively prevented drainage to recharge groundwater. In the absence of such drainage, soil sampling permitted a detailed water balance; comparison with open-water evaporation (Penman estimate) gave annual E_t/E_0 ratios for this closely-grazed pasture of 0·4 in the first year and 0·5 in the second. When fully supplied with water the ratio, from a nearby lysimeter, was 0·75 (Pereira, Hosegood and Dagg 1967).

In a comparison of grass species at greater depth the vigorous Buffel grass (*Cenchrus ciliaris*) depleted the soil profile from field capacity to wilting point to a depth of 6 m, creating a deficit of 610 mm (24 in) eighteen months after planting (Fig. 24). Several common perennial grasses in East Africa have similar rooting characteristics and the creation of deficits of 500 mm or more in very deep porous tropical soils is of substantial hydrological importance. Before recharge of groundwater can occur, the following rains must refill this deficit while meeting current transpiration demands; this requires a total of about 700 mm. For these experiments in Kenya detailed records showed that such conditions had been satisfied only twice in the past sixteen years. Of the total 16 m of rainfall received in these years, only 400 mm would have been transmitted to groundwater by these vigorous perennial pastures (Dagg, Hosegood and McQueen 1967).

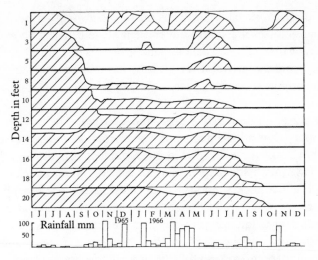

Fig. 24. **Water use by an African pasture grass.**
Soil-moisture tension patterns from resistance of 'gypsum blocks': the shaded areas represent available water at each depth. Depletion is from field capacity to wilting-point throughout a 6 m depth of soil, creating a soil-moisture deficit of 610 mm (24 in), in 18 months after planting *Cenchrus ciliaris*. (Dagg, Hosegood and McQueen 1967.)

Semi-arid grassland

About one-third of the world's total land area is mapped by ecologists as desert and tundra (i.e. hot and cold dry areas respectively) and Dr H. L. Shantz, the veteran specialist in the water relationships of plant communities, estimated that about 6 million km^2 of this area is effectively grassland (Shantz 1954). Hydrologically, some estimate of the lower limit of rainfall at which dry grassland ceases to produce runoff might be inferred from the calculations of Schachori and Michaeli (1965) in a literature review of watershed experiments. By plotting rainfall-to-runoff relationships for published experiments on treeless watersheds (grassland, felled forest, bare ground), they suggested 200 mm as an effective practical minimum at which runoff occurs. In deserts, heavy rainstorms produce local runoff irrespective of the annual totals.

Nomadic populations of graziers have learned to survive in these areas, too dry for crops and beyond the reach of irrigation. Approximately half of Iraq, for example, is in this category, with rainfalls from 350 mm in the best areas to negligible amounts in the true desert. The semi-desert is heavily grazed and browsed by camels, sheep and goats. In a recent survey by Kaul and Thalen (1971) the rangelands are reported as over-exploited and seriously deteriorated. Sparse rainfall occurs as

159

storms of great intensity and runs off to shallow collecting pans. Here evaporation losses are severe but there is some penetration and storage of soil moisture. These shallow pans give effective seasonal grazing if carefully conserved, but over-grazing has reached the stage of soil erosion and has severely reduced carrying capacity.

Range management for watershed control. Such over-grazing is unnecessary, but a high degree of organisation is needed to secure the grazing of such sparse forage without rapid loss of hydrological control. In the USA, with adequate capital for fencing, water development and transport of livestock, semi-arid lands carrying only one beast per square kilometre are successfully grazed in summer, with removal of stock to better range or to fattening yards before the extensive areas are denuded. In Africa, south of the Zambesi River, extensive savanna country, receiving only 300 to 500 mm of rain annually, successfully carries up to five head of cattle per square kilometre on large ranches well equipped with fencing and water-points for rotational grazing. In spite of some thirty years of Commissions of Enquiry into drought no economically viable solution has yet been adopted to meet the periodic failure of the rains which decimate the herds on over-stocked, under-equipped and ineffectively managed ranches, on which the majority of the cattle population are still to be found. The same land-use problems recur in Australian grasslands where the development of highly specialised transport has proved an economic means of moving sheep and cattle between grazing areas and to markets.

The reader may well feel that this is straying from water-resource studies into agriculture, but the vast loads of suspended sediment carried from over-grazed ranchland rapidly fill storage dams and severely limit the irrigation and power potential of rivers critical to the development of community resources. The water-resource authorities need to be actively concerned in any form of range-management improvement which can protect watersheds.

Enquiries in which the writer has taken part point clearly to an integration of the grazing resources with the water resources as the logical and economic solution. The main difficulty of livestock management is that all the forage must be produced in critically short periods of grass growth during the brief seasonal rains. During this growth season grazing does maximum damage. Protein values of the new grass shoots are high, up to 16% or more of the dry matter, but the yield per hectare is negligible. If allowed to grow, a moderate volume of low-quality forage (protein 2–4%) is produced. Not all of this is edible since the lignified (woody) parts of the tall grass stems are not digestible. For survival sheep and cattle graze selectively, licking the edible leaves from the dry stems. Ideally, such rangeland grazing needs supplementing by higher-quality feed and protecting by complete removal of stock during the critical short growing season of the grasses. Even at 300–500 mm there may often be concentration of overland flow at low points and accumulation of groundwater. Local small-scale irrigation schemes to grow forage legumes of high protein content, to provide both keep during the growing season and protein supplement in the dry season, have been recommended. Experimentally, they have been practised successfully on a small scale in areas in which water resources are inadequate for major irrigation schemes. New techniques of water engineering have made possible provision of water-points for rotational grazing on a far wider scale by the laying of cheap polythene piping over distances of several kilometres; provision of water at staging points along offtake routes, construction of excavated tanks for the storage of rainwater from small catchments, and the exploration and development of groundwater supplies by boreholes are all essential contributions to the raising of efficiency of land use and hence of watershed control in dry rangeland.

As in all forms of land use, however, local prevention of runoff leads to more evaporation from the catchment surface and less yield at the final outflow. Culler (1961) reported from a study of the 23,000 km^2 semi-arid basin of the Cheyenne River, Wyoming, a total of 9320 stock-ponds and small reservoirs in the watershed of the Angostura Dam. A thorough study of four years of records of this dam, and of a sample of fifty of the ponds, gave an estimated annual evaporation loss from the whole of the stored water, of 38% of the 6 million m^3 yield of the watershed in a dry year, and 44% of the 22 million m^3 yield in a very wet year.

Sediment flow. As with the forested catchments described in Chapter 4, the most severe transport of eroded soil and rock from semi-arid grassland occurs after major fires. In Arizona an intense accidental fire in 1959 swept over four experimental watersheds after three years of measurement of the yields of water and sediment (Glendening, Pase and Ingebo 1961). The vegetation was dry chaparral rangeland, which provides scrub grazing of over 20,000 km^2 in Arizona. In the three years prior to the fire the cumulative total rainfall was 1700 mm, the runoff totalled 72 mm and the sediment yield was approximately 19 t/km^2 from the immature granitic soils. Within the first two years after the

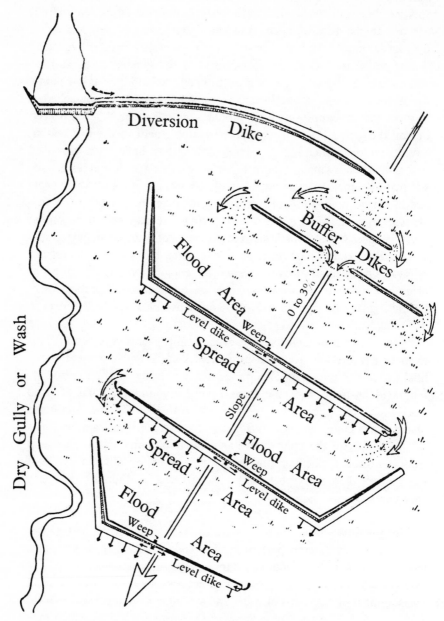

Fig. 25. **Range water spreading system.**
Torrent flow is diverted from small streams and spread over flat valley-bottom land to promote pasture growth. (From Branson 1956.)

fire a total of 1300 mm of rainfall produced 400 mm of runoff, carrying sediment at a rate of 8300 t/km^2. In such semi-arid country the catchment areas of dams therefore need protection from wildfires as part of the watershed management.

The tall stems of the grasses often give an illusion in semi-arid areas of a complete soil cover, an impression which does not survive close inspection. Up to 50% of bare soil may be exposed between tufts of tall, fire-tolerant grasses. These soils may exhibit geological sheet erosion, which continues even when they are completely protected from fire and grazing, e.g. in similar Arizona dry-range country Hadley and Lusby (1967) measured a 50% runoff from an ungrazed basin of 4 ha, with sheet erosion at 37 t/ha from a single 25-mm storm.

Flood-spreading on rangelands. The flash-floods and sediment-transport from dry rangelands can be ameliorated by the techniques of diversion from drainage lines and spreading by contour bunds to improve infiltration. This has been studied on a substantial scale in the western USA in areas of 200–500 mm of annual rainfall. Floodwaters diverted by concrete weirs in the drainage lines were spread by earthern bunds or dykes, 0·6–1 m high. The bunds were made by mechanical equipment, the sides sloping at 1 in 2 and stabilised by sowing with grasses (Miller *et al.* 1969; Plates 16, 17, 18). A typical pattern of spreaders is shown in Fig. 25 (Branson 1956). These devices are particularly effective for flood-spreading over flat valley floors, where hay may be harvested. A great deal of work is involved in developing such spreading systems for large flows. Measurements of the hydrological and pasture effects of a section of the system of flood-spreaders designed to reduce sedimentation of the Angostura Reservoir on the Cheyenne River in Wyoming are described by Hadley *et al.* (1961). Under an annual rainfall of 300–400 mm, mainly as summer storms, a series of 27 small diversion weirs supplied spreaders which distributed flood waters to 146 ha of valley floor. Approximately one-third of the water was retained by the system which trapped three-quarters of the suspended solids. Peak discharges were reduced by two-thirds. In most years the flooded areas produced two cuttings of hay and also winter grazing.

For very dry areas, below 250 mm, the construction of water-spreading systems on rangeland has not proved to yield economic returns. Constructions of such spreaders over several thousand acres, by several different methods, were undertaken by the Civilian Conservation Corps in 1934–42 in Arizona, New Mexico, Montana and Wyoming in order to provide employment. Assessment after twenty

Plate 16. Flood-spreading bunds on rangelands in Montana, USA.
Bunds approximately 1 m high were constructed by mechanical equipment and stabilised by sowing of grasses. They are most useful where 500 mm or more of annual rainfall are received. (From Miller *et al.* 1969.)

Plate 17. Culverts prevent ponding behind flood-spreading bunds.
Water must not be ponded behind bunds for too long, or grass may be damaged by flooding. (From Miller *et al.* 1969.)

Plate 18. Borrow pits provide watering ponds for stock.
Soil is borrowed from below bunds except where a stock-pond is needed.
(From Miller *et al.* 1969.)

years (Peterson and Branson 1962) showed that in very dry areas the effects on vegetation had been slight and were less than the effects of control of grazing.

In wetter areas, with annual precipitation of 500 mm or more, the dykes have to be constructed to a more substantial scale, and ponding is reduced by the inclusion of culverts (Plate 17); the borrow-pits provide storage for stock-watering (Plate 18).

Heat reflection from dry grassland. Both the Sahara and Kalahari deserts are slowly spreading; the desiccating effects of the hot dry winds from the desert on the vegetation of adjacent semi-arid country is the usual explanation, although successive studies have noted that over-grazing of these areas is at least a major contributory cause and may well be the main reason for the slow deterioration.

A study of the physical heat exchange between air and land surface was undertaken on the large cleared areas of the Kongwa cattle ranch in Tanzania (Pereira and McCulloch 1960) by thermograph recording of soil temperatures under tall grass and bare soil. The result was a little unexpected, in that the reflection of heat from the surface of dry grass provided the most effective insulation for the soil surface, there being only a very slight diurnal heating effect. Penetration of the daily heating

165

wave into bare soil was, however, effective to some 15 cm depth, so that there was heat storage during the day and heat emission during the night. More heat was therefore returned at the hottest time of the day from the cover of dry grass than from the bare soil.

Summary of effects of grazing on watersheds

In humid climates on vigorous, well-managed grass swards, water yields increase with the intensity of grazing.

Wetlands yield more water when drained and grazed.

On forested watersheds grazing of domestic livestock must be restricted to glades in which the soil is protected by a grass sward. On slopes where the ground-cover is forest litter, trampling of livestock causes erosion and torrent flow.

On semi-arid rangeland, management of watersheds by control of grazing and of fire is made difficult by characteristically erratic incidence of rainfall and drought.

Concentrations of stock around scarce water points cause severe damage; provision of water supply points is of critical importance.

Development for optimum yields of water and livestock requires organisation of fodder production or conservation to permit resting of rangeland from grazing during the brief seasonal periods of maximum growth of grass.

8

The effects of croplands on water resources

Watershed behaviour

Farmlands under arable crops make a direct demand on water resources and seasonally deplete soil-moisture reserves, but these effects are usually less than those of the natural vegetation which they have replaced. Grains, potatoes, 'roots' and industrial crops will usually have shorter growing seasons and lesser stature than the trees, shrubs, and herbaceous vegetation of the natural ecology, unless this was natural open grassland. From previous chapters it will be clear that the arable crops may therefore be expected in most cases to use less water. Only the irrigated tropical crops, such as sugar and rice, would be likely to use much more water than the natural vegetation which they replace.

The main hydrological effect of arable cropping is on the reception of rainfall and its partition between overland flow and infiltration into the soil. By the exposure of bare soil to rain, sun and wind the capacity to absorb heavy rainfall is reduced, so that immediate overland runoff can produce sharp peaks of streamflow which may combine, in a large watershed, into floods. Cultivation without adequate precautions to prevent surface runoff from reaching erosive velocities can result in transport of surface soil. This produces sheet and gulley erosion of the farmlands with consequent deposition of sediment in channels and storage reservoirs lower in the watershed. Soil in suspension represents a deterioration in water quality which creates serious difficulties in water purification, as the suspended fine particles of clay need expensive treatment by chemical flocculation and filtration. The water-supply engineer thus depends greatly on the success or failure of the farmers in the watershed to manage the air–soil–water pattern within the rooting depth of their crops. In particular the runoff depends on the physical condition in which the farmers maintain their surface soils.

Surface soil management. The practical farmer is sometimes said to 'have his feet on solid ground' but his efficiency depends greatly

167

on his recognition that this is the wrong metaphor. He regards his soil
not as a solid but as a variable assembly of small holes, the chinks and
crevices between the rather fragile crumbs of soil and organic debris.
The larger pore spaces contain both air and water. From the surfaces of
the soil crumbs nutrients diffuse into the soil water and thence into the
roots. Crops therefore flourish only while their roots can freely penetrate this honeycomb (Plate 19), from which they must find air, water
and nutrients (Russell 1972). Excess of water deprives the roots of air,
so that good drainage is important. Compaction by tillage or by the

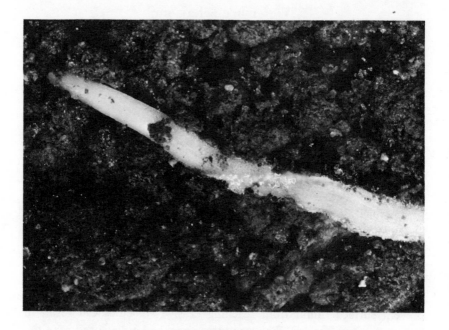

Plate 19. Root penetration through soil pore-space.
Roots, air and water move through the pore-spaces between soil crumbs.
(Photograph by Yoxal Jones from East Malling Root Observation Laboratory.)

passage of equipment while the soil is wet causes the collapse of the
large pore spaces so that the soil locally ceases to become a medium for
root growth. By the timeliness and contour-pattern of tillage and by
growing rotations of well-fertilised vigorous crops, the farmer can exert
a beneficial influence on the hydrological behaviour of the watershed.
The objectives of the farmer and of the water engineer are thus in close
agreement (at least while water is in adequate supply). With arable land,
as with pasture and ranchland, the higher the quality of the agricultural
management the less the hydrological damage. An arable crop with a
168

high density of plant population covers the surface and protects the soil from the kinetic energy of striking raindrops (Duley 1939). Root systems of dense well-fertilised crops contribute many tons per acre of organic matter to the soil. Much direct experiment on soil structure has established that this is improved most rapidly under swards of pasture grasses; 'ley farming', 'alternate husbandry' or 'grass rotations' are terms describing systems in which land carries grass and crops alternatively, each for a few years at a time. It is effective hydrologically and is a basis of much successful mixed farming, but agriculturally it is valid only where livestock can be profitably run. With modern techniques of skilled cropping, on most arable soils, grass is no longer essential for the maintenance of soil structure.

Where soils, topography and climate intensify the hazards of erosion of arable land, the device of 'strip cropping' has been developed by the Soil Conservation Service of the USA. Alternate strips of grass and of cultivation, aligned along the contour, are then managed in an intricate time-table of grazing and of hay cutting which admits the livestock at times when the arable crops will not be damaged. The spectacular air photographs of watersheds under this elegant system are a striking feature of the literature of soil-conservation methods. However, strip cropping is economically viable only where livestock play the dominant role in a mixed farming system, and is not a general solution which can be readily applied in tropical lands.

Tillage practices evolved in Europe under climates of water excess have produced severely adverse effects when exported to climates with long dry seasons and rainstorms of high intensity. In particular the mouldboard plough, which inverts the top 100 to 200 mm of soil in order to bury a grass sod or to bury weeds and weed seeds in an arable field, has been misused. It is an excellent and indeed essential component of farming under high rainfall, where the traditional straight furrows, usually running downslope, aid in drainage. When misapplied in climates in which the exposed soil is desiccated by long dry seasons and is then beaten by rainfalls of high intensity, the exposed bare soil surface slumps into a paste and the downhill furrows contribute to erosive velocities of runoff. Tillage on the contour and the use of subsurface cultivations to leave plant residues as a protective surface cover for the dry season have been the essential modifications, while terracing and the provision of grassed channels for surface drainage at low velocities have been designed as highly adaptable soil-conservation systems suiting a wide range of climates, soils and topography, as illustrated in Chapter 2.

Cropping in climates of seasonal drought

While the very wet and the semi-arid climates present special problems which are considered separately below, the largest areas of developed farmlands for which data are available are in climates having strongly seasonal rainfall with long dry seasons. These can be grouped for hydrological study into climates either of winter rainfall or of summer rainfall, and into tropical climates in which both wet and dry seasons are warm.

The most extensive recording of watershed behaviour under cropping practices in strongly seasonal rainfall has been in the USA under winter rainfall conditions where early and urgent problems were created by mistakes in land use and management. These mistakes, which occurred in the rapid opening up of the virgin lands of a continent, before the problems had been studied, are being repeated in developing countries although the technical solutions are now available.

Soil-conservation effects in the USA. The Soil Conservation Service of the US Department of Agriculture was established in 1935, to provide technical help in development of agriculture with protection of both soils and water resources. In 1957 a valuable series of publications was begun by the Agricultural Research Service of the Department, to collect and put on record the data from watershed studies of agricultural lands. The first volume reported summaries of thirty-four years of precipitation and runoff data from 334 experimental watersheds. These were at sixty locations scattered over twenty-seven States (USDA 1957). This was followed by publication of the thirty-four years of annual maximum flows from the same watersheds (USDA 1958), and then by a tabulation of details of selected storm events from sixty-eight of the watersheds (USDA 1960). All three types of record have been presented simultaneously in subsequent volumes (USDA 1963, 1965 and 1968). The published data are not from experiments in which land use has been changed to observe the hydrological results. They are detailed studies of the precipitation and runoff from land under a wide range of farming practices, and of the way in which small watersheds can, in extreme storm events, contribute to floods. The records include very small intensively studied areas of 0·5 to 3 ha, with a good sample of valley sizes up to 30,000 ha. Individual storm events given in detail include very high flows from small watersheds under crops, an extreme example being a peak flow rate of 152 mm/hr (6 in/hr approx.) from a small 0·5 ha area of maize at Coshocton, Ohio, under a rainfall of 37 mm

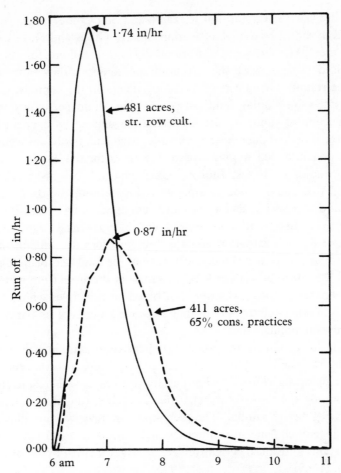

Fig. 26. **Effect of conservation practices on runoff hydrograph at Hastings, Nebraska, from a storm of 2·75 inches.**
Good farming practices on only 65% of a watershed halved the peak rates of runoff experienced under straight-row cereal crops.

at an intensity of 103 mm/hr. Rainstorms of very high intensity are usually localised so that larger valleys do not experience such high intensities over their whole area. The peak flow recorded at Coshocton for the surrounding 1800 ha (7·16 sq. miles) watershed of mixed cropping, pasture and woodland was only 18 mm/hr (0·72 in/hr). From this larger area the volume rate of flow was equivalent to some 90 m³/sec or 3200 cusecs, and the valley shed 50 mm of water in a single day.

The task of extracting from this mass of information some quantitative description of the behaviour of watersheds under different farming practices was attempted by an experienced inter-departmental team from the USDA Agricultural Research and Soil Conservation Services and from the Bureau of Reclamation (Sharp, Gibbs and Owens 1966). After more than five years of work on the data a 'rational formula' was produced to summarise the effects of soil conservation treatments on surface runoff. The effects of terracing, tillage on the contour, seeding of pastures on sloping land, strip cropping, drainage, irrigation and construction of storages were numerically evaluated by curves plotted from the observed data. The areas under each land treatment were then weighted and added to give a predicted effect for the complete watershed. Studies in Texas, Indiana, Oklahoma and Nebraska (Fig. 26) showed that conservation treatments reduced runoff by from 25% to 40% in dry years but had a less consistent effect in wet years. The evidence from direct and statistically valid experiments is, however, inadequate for generalisation and the 'rational formula' treatment was adopted in order to produce early results for land-use planning. Although this form of analysis is of direct use in the areas from which the data were taken, the curves were not based on principles which can be applied to other countries until they too have acquired a similar mass of field measurements.

While it is unlikely that so comprehensive a solution will ever be devised for general application, much practical progress has been made in developing agricultural techniques for preventing soil erosion and for ameliorating the flash-flood behaviour of arable watersheds under intensive regimes of rainfall. The techniques are reviewed and illustrated for a range of climates in an international handbook (FAO 1965). Their applications in tropical agriculture are clearly described by Webster and Wilson (1966).

Practical examples of hydrological improvements. Studies of complete watersheds for several years before and after application of soil-conservation techniques, or experimental comparisons with untreated control valleys, are inevitably rare in a matter which determines the living standards of the farmers concerned. The following three examples illustrate the hydrological results of soil-conservation treatments.

The Tennessee Valley Authority's grand design to develop the natural resources of a watershed of 100,000 km^2 has been quoted in Chapter 4, where the very substantial hydrological results of reafforestation were summarised. Another TVA experimental study was made on the Parker Branch watershed of 43 ha under mixed farming: annual precipitation is 1000 mm. At the beginning of the study the land was in a seriously deteriorated state with 30% of the surface area gullied and abandoned. Simple improvement of farming practices and application of conservation rules led, over ten years, to crop increases of 60% in maize, 50% in tobacco, 25% in wheat and 117% in alfalfa (lucerne) hay. The reduction in peak rates of stormflow in Parker Branch Creek are illustrated in Fig. 27. These reductions were from storms falling on a dry catchment

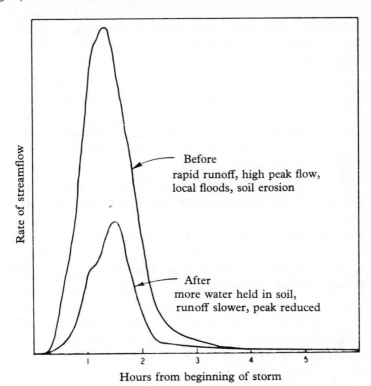

Fig. 27. **Hydrological effect of improved mixed farming on an eroded watershed.**

Flow records from Parker Branch experimental watershed under the Tennessee Valley Authority. These illustrate the results of soil conservation practices and improvements in the mixed farming of an eroded catchment area.

area. The average peak flows from summer storms were halved. Where the land surface was fully wetted, stream patterns showed no effective change. Annual soil transport was, however, reduced by two-thirds, from 7·5 to 2·5 tonnes/ha/yr of suspended sediment. Less of the water ran over the surface and more penetrated the soil, but the changes were rather small; estimates were 3% reduction in surface runoff, 2% increase in groundwater and 1% increase in water yield. Differences of this magnitude are difficult to establish experimentally and the main effects were the improvement in water quality and the control of summer storms (TVA 1963).

A striking example of improvement in a city's water supply by better arable farming on the watershed is that from Atlanta in Georgia. The watershed of the Chattahoochee River covers 370,000 ha of foothills and plateau, about half under forest, but with the rest in small farms which, in 1931–5, were under a poor standard of subsistence agriculture. Analysis of turbidity measurements over twenty-five years showed 500 to 800 ppm of suspended sediments in the summer flows from 1931 to 1937. Closing of the steeper land to cultivation by organising forest reservations affected only one-twentieth of the catchment area, but the agricultural practices on the whole of the arable land were improved and over 6000 km of terraces were built. There was a steady improvement in the water quality and from 1950 onwards turbidity has been less than 100 ppm (Albert and Spector 1955). These studies of the earlier and more severe soil-conservation and water-supply problems of the USA are important today because many developing countries now face similar situations.

A third example is from the Soil Conservation Experiment Station at Coshocton, Ohio. There, two watersheds under unimproved 'prevailing practices with average crop production and average soil stability' were compared with two others under full soil-conservation practices, 'high crop production and considerable soil stability'. Results have been reported from 1938 to 1968. The general conclusion has been that the soil-conservation practices decreased most of the stormflow peaks although exceptional very large peaks were not reduced. The improved farming practices increased infiltration of rainwater and decreased stormflow. These effects increased over the first two decades of the experiments but there have been no significant changes in the past ten years. Lysimeter studies described in Chapter 3 indicated these changes as due to deeper root ranges and greater water use of the more vigorous crops (Harrold *et al.* 1962; Ricca *et al.* 1970).

Surveying the land-use situation in the USA after some thirty-five years of Federal and State campaigns of soil conservation, Robinson (1971) concluded that the objective of reducing soil loss to 2·5 tons/ha per annum is far from being achieved. The average for the South-East is still about 15 tons and for the Mississippi lands the erosion loss is from 20 to 30 tons/ha per annum. The Mississippi River alone carries 500 million tons of sediments into the Gulf of Mexico annually. This flow is estimated to contain some 17 million tons of plant nutrients.

174

Thus even in a technologically advanced nation, the need for better soil conservation remains acute.

Part of the obstacle to better soil conservation is that terracing and tillage along the contour cause operational difficulties with agricultural machinery, which works more economically in rectangular fields. Although the benefits are real to the downstream community, and ultimately to the long-term profitability of the individual farm, the immediate and visible benefits may be too slight to outweigh the nuisance of conservation measures. In the Middle West of the USA, while driving around a contour-tilled landscape with the head of the local Soil Conservation Service, the author was told that, initially, the local campaign to introduce conservation methods had met with no response. Eventually the parson was convinced and a fiery sermon on the duty of the husbandman to the next generation who must till the land after him, triggered a general response.

In addition to watershed studies there has been an abundance of reports, over many years, of measurements of the runoff from sloping plots of land under various experimental crops and tillage treatments. These have contributed to the understanding of the processes of infiltration and surface flow, but their results are difficult to relate quantitatively to natural watersheds.

Experiments in the USSR. There are many reports of such studies from the USSR, in addition to the streamflow studies already referred to in Chapter 4, but the techniques of soil conservation do not yet appear to have been strongly emphasised in agricultural policy. Data from many plot experiments have been surveyed by Unofrienkov of the Ukraine Institute of Hydrometeorology at Kiev, where the writer had the opportunity to discuss the work on site. Bochkov (1963), from the State Hydrological Research Institute of Leningrad, has also collected and analysed runoff data for a wide range of conditions, many of them dominated by the effects of soil freezing and snow melting.

Summer rainfall and winter drought. Transfer of soil-conservation techniques to areas such as Southern Africa and Northern Australia in which there are dry winters has needed some substantial modifications. The general policy of leaving stubbles and plant trash on the surface, to protect the soil from the direct impact of rainfall, has relied in North America on the rapid weathering of the plant material in a wet winter, so that it is readily pulverised by simple machinery and does not seriously interfere with cultivations in the spring. Attempts to use these practices when the winter is dry, particularly in Africa where the tall stems of maize or sorghum are ubiquitous, result in an embarrassing residue of hard dry stalks which impede both planting and weeding when the summer rains begin. This is met by the use of heavy disc ploughs or by rotary tillage, and also by removing the stubbles either by grazing or by burning. The increased use of herbicides instead of inter-row weeding may effectively change the cultivation requirements in intensive farming under such climates, and improve infiltration of rainfall.

The main hydrological difference produced by summer rainfall, however, is the increased vulnerability of the water supply to evaporation losses from a frequently wetted landscape in warm summer temperatures. The interests of the water engineer and the farmer remain identical so far as quality of water is concerned, but diverge on the issue of quantity. By his aim of retaining water where it falls the farmer contributes the minimum to the filling of reservoirs. Deep-rooted crops create soil-moisture deficits which absorb summer rainfall. The ultimate solution by the surface-sealing of special catchment areas has been discussed in Chapter 2. These problems are shared by the high-altitude tropical areas which experience similar conditions for arable agriculture.

Tropical problems. In the developing countries the difficulties have been discussed in Chapter 2. The problems of improving land use in tropical watersheds of developing countries, before erosion destroys too critical a proportion of their soil and water resources, remains one of the major large-scale problems facing mankind.

Weak, tropical kaolinitic clays from which Pluvial Era leaching has removed the silica, are subject to rainstorms of destructive intensity. Planting on ridges is an indigenous tradition in many such countries, but the challenge is to teach the techniques of constructing these ridges along the contour.

Such contour ridging can be made more safe by 'tying' the ridges, sometimes called 'box-ridging'. Adaptation of modern tillage equipment (Farbrother 1960) has made it possible to drag the loose soil between the ridges into small dams or 'ties' (Plates 20, 21) to form rectangular troughs which hold the rain until it has time to soak in (Cashmore and Hawkins 1957; Hawkins 1959; Lee, Ofield and Passmore 1960). This is more effective, in tropical soils, than the earlier US technique of 'basin-listing', or scooping out a succession of hollows for collecting water. The tied-ridging is very effective in controlling soil movement

176

on land with slopes up to 12% in red tropical kaolinitic clays (Pereira, Hosegood and Dagg 1963; Dagg and Macartney 1968).

A major difficulty in tropical areas of subsistence agriculture is that rapidly increasing human populations have spread from the valley floors up the steeper hillsides on to slopes of 20% or more. In the extreme cases where growing populations are surrounded by steep hillsides, as in parts of Uganda and the ex-Belgian Congo, hoe cultivation

Plate 20. Tying of ridges to improve rainfall penetration.
Co-operative work between the National Institute of Agricultural Engineering and agricultural research stations in East Africa demonstrated the success of this simple device for building small dams in each furrow. The trailing discs are raised by eccentric land-wheels. (Photograph by National Institute of Agricultural Engineering, Silsoe, UK.)

has been continued to the stage at which cultivators were belayed by ropes to avoid falling out of their fields. In South-East Asia and in the old Inca traditions of South America very narrow irrigation terraces have succeeded in conserving both soil and water on steep slopes, but such techniques are rare. More usually lands are ploughed wherever the oxen can climb, and, when side slopes are difficult, they tend to work straight downhill, e.g. the photographs in Plates 22 and 23 were taken by the author in Ethiopia recently. The soil-conservation tech-

Plate 21. Holding rain on the land where it falls.

Tying of ridges improves rainfall penetration, increases crop growth and decreases soil erosion. These beneficial hydrological effects are accompanied by increase of water loss by evaporation and by transpiration, with a consequent reduction in the water yield of the catchment area. (Photograph by National Institute of Agricultural Engineering, Silsoe, UK.)

Plate 22. Ploughing downhill in semi-arid land may help the oxen but also accelerates soil erosion.

Cultivation on the contour, the use of fertilisers, and seed of improved varieties of cereals, are all needed to raise the income of these hard-working farmers. A soil conservation service has newly begun in Ethiopia.

Plate 23. The result of ploughing unprotected slopes.
On steep slopes above roads, the roots of trees, shrubs and grasses bind the soil. Clearing and ploughing are followed by torrent flow and gulley cutting.

niques worked out for sound agriculture would usually leave such steep slopes under grass or trees. On gentle slopes the terraces which limit runoff velocities in arable farms in the more developed systems are broad low banks and shallow ditches over which tractors and even combine harvesters can pass without difficulty. A high priority for the tropical watershed engineer must therefore be to persuade his agricultural colleagues to keep their ploughs and hoes away from the steeper hillsides and to apply the proven techniques of soil conservation to the lesser slopes.

A critical limiting factor for such progress is the teaching of simple techniques which the farmer can understand and use. An example, from personal experience, is the task of setting out the contour banks and furrows. As developed by practical agricultural engineers of the former British services in Kenya, this can be done with the simplest 'line level' equipment (Layzell 1968). With two sticks, 10 metres of string, and a carpenter's spirit level, the writer has been able to give one morning's instruction to illiterate labourers and to leave them to set out contour banks on 3 hectares of hillside; the resulting banks and furrows subsequently survived the searching tests of tropical rainstorms. Yet in other developing countries, visiting international experts, familiar only with sophisticated techniques of theodolite and bulldozer, have advocated the university training of engineers in these skills as the first step. The trained graduates go off to administrative jobs while the soil erosion on the hillsides continues its relentless destruction of prosperity.

As the Incas and their descendants have successfully demonstrated with their irrigation systems, soil and water management requires simple skills and discipline rather than complex and expensive techniques.

Cropping in climates of water excess

In Chapter 3 it was noted that Penman (1950a) has accounted for the annual water balance of the main catchment areas of England by the assumption that the whole land surface is a continuously transpiring canopy of vegetation. Indeed from the air this assumption can be subjectively confirmed, although seasonal deficits are sufficient in the south-east for crops to respond profitably to supplementary irrigation.

As a result of mild and well-distributed rainfall, exposure of bare soil is rarely the cause of erosion and there is rapid restoration of a cover of crops, grasses or weeds. Soil erosion and suspended sediment transport are therefore usually minor problems and farming is maintained at a generally high standard. The main result of the conversion of grassland or woodland to arable land under these conditions is to produce a more rapid drainage response and a small net increase in water yield.

Tropical climates of water excess at high temperatures favour the development of tall dense forests which collect and hold a precarious supply of nutrients. While the Ice Ages protected high-latitude soils, the associated Pluvial Era leaching in the tropical latitudes removed much of the soluble nutrients. When the forests are cleared, arable agriculture involves a struggle to maintain adequate plant nutrition in a climate of frequent leaching. The system of shifting cultivation developed by indigenous tribal communities left most of the land area under protective vegetation and was of little hydrological consequence while population density was low. With rapidly growing populations resting periods of land are reduced and levels of fertility have fallen. Starved crops give poor soil protection from tropical rainstorms and soil erosion is characteristic of high-rainfall tropical areas of high population density.

The International Institute of Tropical Agriculture, developed initially by the Ford and Rockefeller Foundations, and now supported by a world organisation of donor agencies, has been newly established in Nigeria to study this problem (IITA 1971).

The hydrological effects of clearing such rain forests, and some experimental evidence for the quantitative hydrological results, were quoted in Chapter 6. Vast areas of rain forest, as yet undamaged by man's activities, remain in South America, West Africa and South-East Asia. When they are scheduled for development, as appears to be inevitable under man's present population growth, they should preferably go to tree crops, cocoa, coffee, tea, rubber, etc., rather than to arable use. Where they must be used for food crops these will maximise the loss of the regulating effects of the forests on streamflow, and the effects should be studied on pilot-scale watersheds with simple measurements of rainfall and streamflow. The necessary flood-delaying structures can then be designed and installed before clearing of major areas can produce embarrassing floods. Such pilot watershed studies should have a high priority for international aid, and the time to undertake them is during the long process of evaluation missions and the search for international capital, rather than after development has begun.

Semi-arid croplands

All of the destructive processes of soil erosion already discussed tend to be exaggerated when marginal areas of low rainfall are cultivated for field crops. The reliability of rainfall tends to decrease sharply with the annual totals so that at the lower rainfall limits for crop growth the average annual rainfall has little meaning. It is the rainfall which can be expected to fall in at least four years out of five which determines both the local vegetation pattern and the value of the land for cropping. Glover, Robinson and Henderson (1954) plotted the rainfall reliability of East Africa in these terms and showed a close correlation between human population density and the expectation of receiving at least 20 inches (500 mm) in four years out of five, which is the rainfall necessary for a sorghum crop.

A mechanism for survival. Sorghums, millets and maize are often sown in dry soil when the rains are expected, in order to take advantage of all the rainfall, and also of the release of nitrogen from the clay particles on first wetting (Birch 1960). Such early planted grain often has to endure dry spells. The way in which slight rainfalls of a few millimetres, which are rapidly lost by evaporation from the wet soil surface, can become concentrated around the roots of the individual stems of grain-crops, was demonstrated by Glover and Gwynne (1962). They collected flows of about 500 ml from a single stem of droughted corn in a 4 mm shower, and showed by excavation that a hemisphere of moist soil could be found at the bases of such stems after light showers, in the zone in which roots are plentifully developed. Over-year storage of soil moisture also can provide viable crops with only 200 mm annual rainfall (Pereira *et al.* 1958), when weeds are controlled.

The penalties of misuse. Few quantitative watershed studies of semi-arid country under such marginal agriculture are reported. One such

Plate 24. Restoration of soil stability on steep land in Algeria.
This thorough terracing was photographed four years after construction; the object is revegetation and eventual afforestation to protect roads and farmlands below. (Service des Eaux et Forêts d'Oran.)

study, carried out by the French colonial agricultural engineers in Algeria (Saccardy 1959) reports land misuse by nomadic tribes, with population increases resulting in the 'relentless cultivation' of wheat and barley. The rainfall intensities quoted are typical of those of East and Central Africa, with rates of 75 to 150 mm/hr falling on bare soils or sparse, droughted crops, cultivated without regard to soil or water conservation. The damage is progressive, and over the 1950–7 period damaging floods in the Oran, Bone and Phillipville areas of Algeria increased annually. The maximum recorded loads of suspended sediment rose from 15,000 to 100,000 ppm and one catchment surveyed in detail

182

(la cuvette du Barrage de l'Oued Fodda) was losing 3·5 million tonnes per year, or about 7 mm over the 75,000 hectares. The remedy proposed was full terracing to hold the rain where it fell and photographs were given of several square kilometres of completed terraces. Plate 24 shows steep land terraced for revegetation to protect roads and farmlands below (Plantie 1961). Labour for both construction and maintenance, although genuinely necessary if high population densities are to subsist by cultivation in such a harsh environment, is costly in comparison with the crop yields achieved, and would be sustainable only by increasing the agricultural inputs.

The alternative, if the population pressure continues without soil protection, is the complete destruction of the agricultural potential of the area.

Protection of soil and water on arable lands

The arable croplands carry the major burden of sustaining human food needs. The difficulties of farming them increase with increasing temperature and the declining amount and reliability of rainfall. When well farmed they yield water free from suspended soil. Techniques of farming with conservation of soil and water resources are well known, but they need community organisation, technical instruction and the capital inputs for fertilisers, seeds and services. As the standard of farming improves in summer-rainfall and semi-arid climates, runoff decreases. Where the standard of farming deteriorates to the stage of soil erosion, water yield increases in the form of stormflow, bearing increasing quantities of soil in suspension. The interests of the farmer and the water engineer are thus inextricably interlocked.

9

The roles of irrigation and drainage in water resources

We are not concerned here with the techniques of irrigation or of drainage, but with their hydrological effects. Irrigation systems convert dry soil surfaces into transpiring crop canopies over areas of the order of 26 million ha in India, 15 million ha in the USA and 12 million ha in the USSR; they are clearly major consumers of water by evaporation. Efficiency of irrigation is therefore important to water resources on a world scale.

A general result of research in recent years has been to show a low level of efficiency in water use to be a characteristic not only of traditional peasant areas of irrigation but also of modern schemes designed by qualified engineers. A comprehensive monograph on modern irrigation has been published recently by the American Society of Agronomy (Hagan, Haise and Edminster 1967). Exchange of information on technical advances in these fields is organised through the International Commission for Irrigation and Drainage (ICID) from a permanent headquarters in New Delhi. A major international conference is organised by ICID every four years; the Proceedings of these conferences are supplemented by annual bulletins and bibliographies which summarise the professional meetings of Civil Engineers in many countries. This literature shows an increasing awareness of the long-term hydrological effects of irrigation schemes.

Groundwater and salinity

The main problems, additional to the direct water use, are those of drainage and the control of salinity.

Irrigation is most rewarding where solar energy and favourable temperatures support vigorous crop growth, so that the main developments are in warm climates with long dry seasons. Supply of water in such circumstances receives far more attention than does the removal of accidental water surpluses. In the level topography most suitable for irrigation, a combination of canal seepage and the over-watering which is a universal tendency of farmers, can raise groundwater levels over large areas.

This has little effect until the groundwater reaches a critical depth of about 1·5 m from the surface. At this stage the 'capillary fringe' of water films, extending upwards from the water table, reaches the high-temperature zone of rapid evaporation in the top few millimetres of soil. Dissolved salts, left behind in the soil as the water evaporates, concentrate near the surface and severely reduce crop growth.

The flat areas of alluvial soil favoured for irrigation are usually well supplied with soluble salts, but while drainage is adequate, excess salts can be leached down by irrigation and disposed of in the drainage water.

With inadequate drainage, however, leaching with excess water merely accelerates the rise of the water table.

The 4000 years of written records of the irrigation systems on the Tigris, from 2600 BC to 1400 AD, have been studied jointly by the Director-General of Antiquities in Iraq and the Oriental Institute of the University of Chicago (Jacobsen 1958). They include a series of recorded disasters from salinisation. Salinity surveys are recorded from about 2400 BC and control by leaching and drainage appears to have been practised. As mentioned in Chapter 2, however, the continued success of irrigation schemes appears to be very closely bound in history to the existence of stable and vigorous central government. As Jacobsen points out, it is necessary for government control to extend beyond the irrigation areas to the protection of the watersheds.

In modern times there has even been some increase in the hazard, as political priorities for capital investment have tended, for prestige reasons, to support new storage and command development rather than the renovation of declining areas. The losses from waterlogging and salinity therefore continued; only a decade ago it is probable that a larger area was lost to cultivation every year from salinity than was gained by new irrigation schemes.

A widely successful solution has been to pump from the groundwater. Campaigns for the installation of simple tube wells, with local manufacture of equipment, are now reversing the trend of salinity losses in Asia over important areas where the salt content of the groundwater is low and only the concentration of salt by capillary rise from a high water table causes salinity damage to crops.

This encouraging development is an unspectacular but essential element in the 'green revolution' in cereals production in Asia. Improved cereal varieties came initially from the International Institutes, for maize and wheat (CIMMYT) in Mexico and for rice (IRRI) in the Philippines, funded by the Rockefeller and Ford Foundations and by USAID. These International Institutes have been distributing their new high-yielding varieties, not only for immediate use, but also for adaptation to local conditions by crossing with the best varieties of regional and national plant-breeding programmes in the Middle East, Asia and South America. With the improved varieties, the vigorous 'outreach programs' of the International Institutes have stimulated both national government and international support for 'package deals' by which the new seeds are supplied with fertilisers, pesticides and with improved regimes of irrigation and drainage.

The hydrological results are parallel to those described in Chapters 7 and 8 for the control of soil erosion and torrent flow in over-grazed and primitively tilled watersheds. Hydrological difficulties caused by land misuse are corrected progressively as agriculture emerges from subsistence to cash-economy levels of production.

The political danger still remains, and is indeed similar to that which still affects the provision of sewage disposal plants in the more industrial countries. The allocation of capital to the immediately rewarding schemes of impounding and distributing water has more popular appeal than has expenditure on costly drainage systems to ensure the future.

A practical example. A recent study of major modern irrigation schemes in the Punjab shows that seepage from unlined canals has, in the first ten years of operation, raised the water table 7 to 9 m above the long-term levels recorded since 1895. The Bhakra Canal Area commands 2·71 million ha (6·7 million acres). Waterlogging and salination have already begun as a result of seepage from the 4800 km of canals, which at full supply carry 353 m³/sec (12,500 cusecs). Jatindra Singh, Gajinder Singh and Rattan Singh (1967) report in the ICID Bulletin that drainage on a substantial scale is urgently needed in the Bhakra Canal Area.

Effects on major rivers

Since rivers are often used both as water sources and as drainage outlets, a succession of irrigation schemes has the progressive effect of reduction of the flow in the river with increase in its salinity.

The River Volga. An unusually large-scale example of this effect is causing serious concern in the USSR, where the River Volga is exhibiting these effects (Rutskovskaia 1959). The Volga is the main source of supply for the land-locked Caspian Sea, which is surrounded by semi-arid country. The Volga basin has an area of 3·7 million km² while the Caspian Sea has about ten times this area from which to lose water by evaporation. The hydrological balance is therefore very susceptible to disturbance by large-scale, unplanned land-use changes in the catchment basin. The watershed was once heavily forested, but since about 1600 AD the forest has been progressively destroyed, and much of the farmland has been developed by irrigation from the river. In the lower Volga basin there are many large-scale irrigation schemes and also very many smaller *ad hoc* operations which are estimated to add some 66% to the planned abstractions. The total quantity of water abstracted was 7·68 km³ in 1950, of which approximately 6 km³ were not returned. By 1960 the total abstractions were 15·27 km³ of which some 11 km³ were lost. There was thus an increase in net irrigation loss of more than 80% in ten years. Higher in the watershed, new reservoirs for power and irrigation have added to the area of evaporation losses. As a result there has been a continuous fall in the levels of the Caspian Sea, which have dropped 2 m in the decade 1961–71, while the value of the very important fisheries harvest has also fallen heavily. A recent report (Kirby 1972) ascribes the declining catches of the more valuable species of fish both to industrial pollution and to the interruption of spawning routes by the hydro-electric developments at Volgograd (formerly Stalingrad).

The Murray River. A well-documented example of the acutely adverse hydrological effects of irrigation on a major river system is provided by the Murray River in Eastern Australia. In this case the problem is experienced by a modern commercial farming community, free from the weight of peasant traditions of irrigation. The Murray River system drains one-seventh of the Australian continent and has an average annual yield of some 15 km³ (12 million acre-ft). The surface soils are not saline but the irrigation schemes very soon experienced local salinity problems and within twenty years highly saline groundwater had risen generally to within a few feet of the surface. The salinity problems are described in detail by the State Rivers and Water Supply Commission of Victoria (Currey, Webster and Macleod 1967). Although the sea invaded the basin at the end of the Miocene Age and deposited some 300 ft of saline alluvium, the present salinity problems are due to successive depositions of salt on subsequent land surfaces during periods of aridity.

There are no reliable records on the depth of the water table at the beginning of the irrigation schemes, but it is believed to have lain between 6 and 9 m below the surface. No adverse effects were observed until the 'capillary fringe' of the rising water table reached the soil surface. In the Murray basin this critical stage occurs at a water-table depth of about 1·2 m. Experiments have already shown that no forms of surface drainage or of soil treatment are then effective until the water table is lowered. However, in spite of this critical situation, there appears to be little direct effort to encourage farmers to use water more economically. Applications are controlled by engineering considerations of delivery rates and block schedules; water is metered to individual farms, but charges are low.

The saline groundwater prevents the simple solution of tube-well irrigation which is dealing successfully with similar problems in Asia. Groundwater pumps are indeed being operated with success, but the effluent has almost the salinity of sea water, containing about 3% of salts. A pump, yielding approximately 20 m³/hr, lowers the water table to about 1·5 m over some 50 ha at a capital cost of $25/ha and a running cost of $5/ha/yr. The salt water must be discharged clear of the irrigation area. The Murray River would be the cheapest outlet for disposal, but the pump drainage of the areas at present requiring treatment would add some 300,000 tons of salt annually at a point where the river flow is only 17 m³/sec (600 cusecs). Costly schemes are now in progress to pump the saline drainage water, at critical times of low river flow, to shallow lakes which act as evaporation pans. The pump reclamation of these irrigated lands is thus directly limited by the availability of disposal routes for the saline water.

This severe water-resource problem is made worse by the natural drainage, which, fed by the irrigation waters, already adds embarrassing quantities of salt to the Murray River from some of the irrigation areas on its tributaries. In 1966–7 the Sunraysia discharged 1·1 million m³ at 2400 ppm salt content, or about 53,000 tons of salts into the main river.

The base-flow from the Barr Creek carries 400 tons of salts per day into the river, increasing to several thousand tons per day when rainstorms follow periods of dry weather and wash accumulated salt from the soil surface of the irrigated areas. The problems of communities dependent on the lower reaches of the Murray River for their water supplies are thus formidable. The river serves three States but is operated integrally by the Murray River Commission which guarantees to South Australia minimum quantities of water at flow rates designed to limit salinity to tolerable levels for irrigation and also for supplementary urban supplies for the city of Adelaide.

188

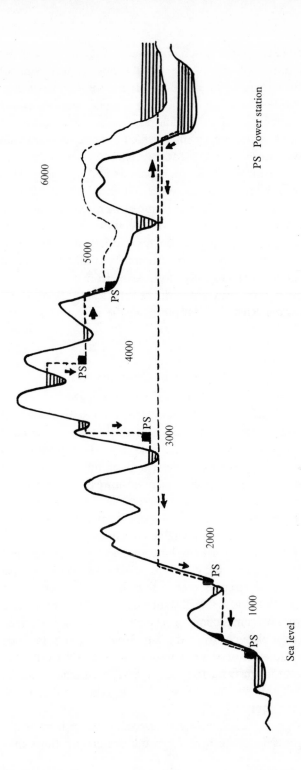

Fig. 28. **Water-resource development in Australia.**

The Snowy Mountains Hydro-electric and Irrigation Scheme collects water from 5000 km² of mountainous country. There are 16 large dams, 160 km of tunnels and seven power stations. The planned peak load is 4 million kW and the annual addition of irrigation water of good quality to the Murray and Murrumbidgee Rivers totals some 2·5 km³ (2 million acre-ft).

PS Power station

6000

5000

4000

3000

2000

1000

Sea level

PS

PS

PS

PS

PS

PS

This need for more irrigation water of higher quality in the Murray River system has produced one of the world's most enterprising of water-resource developments. The Snowy Mountains Hydro-electric Scheme, already mentioned in Chapter 2, has trapped the flow of the largest of the steep mountain rivers which pour eastwards into the sea and has led the water back through tunnels which traverse the mountain range to deliver westwards into the Murray River system (Fig. 28). Sixteen major dams are linked by 160 km of tunnels and by over 600 km of aqueducts. Seven hydro-electric power-stations have a planned output of nearly 3 million kW. Water is pumped from one reservoir to another to give a remarkable flexibility of operation. Extra reservoir storage capacity provides flood protection for the irrigation schemes. This vast development contributes 2500 million m^3 (2 million acre-ft) of good-quality irrigation water annually. By release from the reservoirs during the dry season this clean water dilutes the saline flows which enter the lower Murray River by seepage from the irrigation systems (Hudson 1971).

Effects of over-pumping of groundwater

The advantages of groundwater storage have been reviewed in earlier chapters. The hydrological effects of those irrigation schemes which rely entirely on groundwater are determined by the balance of the rates of pumping and the rates of recharge of the aquifer. In the absence of legal powers to control exploitation excessive pumping has occurred, as in the Salt River Valley in Arizona where an unrestricted development of boreholes in the cotton fields lowered the water table some 100 ft in ten years. The extra cost of power for pumping is some deterrent, but borehole failure is the inevitable endpoint.

While such overpumping can usually be corrected by legislation and control, two permanently damaging effects on water resources may arise in particular geographical circumstances. The first is the permanent reduction of the capacity of the aquifer by subsidence of the overburden. This is ascribed to the compaction of compressible clay materials. Poland (1960) gives an example from the San Joaquin Valley in California, where a net amount of some 4 million acre-ft (5×10^9 m^3) of groundwater was withdrawn from 1943 to 1953. The volume of subsidence was about half this amount, thus seriously reducing the future supply capacity of the aquifer.

The second damaging circumstance is where saline water is normally held back by the pressure of fresh underground water, but flows into the aquifer when fresh water is pumped below the danger level. An example from irrigation schemes on the coastal plains of Israel has been quoted in Chapter 1.

Efficient use of water for irrigation

Three developments in irrigation techniques, which are currently under research in many countries, may be expected to produce more crops with less fresh water on a world scale. These are, firstly, studies on the lining of canals and channels to prevent seepage losses; secondly, developments of crop selection and soil- and water-management techniques to make more use of brackish water; and, thirdly, development of systems of monitoring and control of water distribution to accord more closely with the optimum water requirements of crops. The scientific understanding of plant water use, reviewed in Chapter 3, has progressed far beyond the practical techniques of irrigation. The main priority at present, from the viewpoint of water-resource conservation, is not for more research on water use of crop plants. This, as a convenient university exercise, is now being studied in very great detail in many countries. The priority need is for research and development of relatively simple and inexpensive systems of measurement and control of water application in the field in response to plant need. This will eventually lead to substantial modifications in the engineering design of supply systems to permit greater flexibility of application.

Biological hazards of irrigation

Unwelcome invasions by both plant and disease organisms have proved even more difficult to remedy than many of the salinity problems already outlined. A sombre aspect of the valuable contribution of irrigation techniques to world food supplies is the increase in the incidence of bilharziasis in the human population. This debilitating disease is caused by a minute parasitic worm which spends part of its life cycle in a common tropical species of water snail and part in the human body. The disease organism is not introduced by irrigation; it is usually endemic. From the human population it circulates, by urination, into the snails of existing streams, ponds and marshes and thence by water contact back to humans. Newly-created irrigation schemes may be free of the disease for the first few years, especially if snail populations are suppressed by chemical means. Copper sulphate is often used, but safer and more effective compounds are available, although expensive. A recent isolation of a snail-killing compound from a common African plant, *Phytolacca decandra*, may prove a more economical means of protection for extensive schemes. The reed-beds along the vast shore-lines of the major new reservoirs provide meeting places for snails and human populations. The latter fish, bathe, wash their clothes and draw their water supplies in ignorance of the disease, and indeed they are usually indifferent to warnings of the dangers. Literacy does not seem to make the warnings more effective. A recent survey in Rhodesia showed over 70% of white school children to be infected, in spite of public health campaigns. At present there seems all too little prospect of control of this disease.

Invasion of developed water resources by rapidly growing water weeds provides a parallel set of problems. The water hyacinth *Eichhornia crassipes* and the water fern *Salvinia auriculata* are the chief offenders. In the reservoirs and channels of irrigation systems these can be controlled by herbicides, but on very large lakes and rivers the costs of treatment are prohibitive. *Salvinia* threatened to eliminate fishing and navigation on the 175-mile Lake Kariba while it was filling, but fortunately the weed is damaged by wave action, and has been unable to invade the major open-water surface, although it remains a nuisance in creeks and harbours.

Water harvesting for supplementary irrigation

While the large-scale schemes discussed above exert the most direct influence on water resources, a balanced picture must include the count-less local schemes whereby subsistence cultivators in semi-arid climates collect and store water running off from slopes and use it to improve crops on a small part of the area. These devices are as old as human history, and many tanks and storage ponds of great age remain in use today. Others, as mentioned in Chapter 2, have lapsed through the failure of community discipline but are still viable when renovated. The work of the team led by Professor Evanari in the Negev Desert of Israel was referred to briefly in Chapter 2. This is a determined attempt to understand how the Nebataean peoples survived some 2000 years ago in rugged desert country receiving an average of only 100 mm (4 in) of annual rainfall. The furrows which collect runoff water from bare rocky hillsides have been repaired and the small gardens replanted. The water flows and soil-moisture storage have been measured using modern techniques, including neutron moisture meters. The most suc-cessful ratio of catchment basin to cropped area is reported as about

192

30 to 1. High yields of cereals, fruit and forage (*Atriplex* spp.) are harvested from the very small basins (Evanari, Shanan and Tadmore 1968; Shanan *et al.* 1970).

Small-scale water harvesting techniques are being studied afresh in India today, where village 'tanks', draining several hundred hectares of village lands, and storing enough water to irrigate a few hectares of cereals, are shared by the village community. Deepening, sealing against seepage, and provision of siphons or pumps for irrigation, can greatly improve the efficiency of such water storages, which are made annually more necessary by a rising population. These storages are of particular interest on the extensive areas of heavy black clays, where, in spite of a rainfall of 1000 mm, there is poor penetration into the soil and a 40% runoff, which contributes to floods rather than to crops.

The advantages of storing such harvested water underground have long been known, empirically, to subsistence cultivators in the Middle East, where low stone walls built across the sandy flats at the mouths of 'wadis' check and spread the occasional torrents, so that water sinks into the alluvial soil. Barley is often grown on the stored soil moisture. Crops are best grown where fairly coarse sands provide good aeration, as in the Gash Delta in the Sudan where cotton crops have been grown for many generations entirely on the water from floods held in the soil, into which it penetrates as deep as 5 m in places (Tothill 1948).

When heavy soils are recharged by flooding, insufficient air space is available for roots to penetrate freely. Cereals therefore grow well in the vegetative phase of active root extension, the roots growing into the zone of aeration as they use up the available water. When the grain begins to fill, however, the supply of materials assembled by photosynthesis in the leaves is diverted from root growth to seed growth; root growth almost ceases, the water available in the volume of soil exploited by the roots is rapidly exhausted and the crop dies unless further irrigated, making no use of the water beyond the root tips (e.g. Pereira 1958). Recharge of groundwater by spreading of seasonal floods, followed by irrigation from shallow wells, therefore offers a productive use of torrent flows which would otherwise be wasted.

An enterprising modern example of water-harvesting, with a minimum of capital expenditure, was seen recently by the writer in the desert country on the coast of the Eritrean province of Ethiopia. Here torrents rush annually from the steep and rugged escarpment to escape to the sea along wide 'wadis' which are dry for most of the year. A combination of Italian and Ethiopian enterprise brings a bulldozer every dry season to spend three days in piling up a barrage of sandy soil across the bed of the 'wadi' and in cutting a diversion ditch to lead the water into alluvial flats nearby. The first few floods of the season are thus diverted to soak an area of about 25 ha of desert. As the watersheds become saturated, the torrents increase and the barrier is washed away. This is desirable,

Plate 25. Salad crops in the desert.

Diversion of annual floods in the main channel to recharge groundwater on alluvial flats. The crops of tomatoes, cucumbers, peppers and egg-plants are watered from shallow wells drawing on the water stored from the flood. This enterprise was observed in the coastal plain below Asmara.

since the developed area would otherwise be overwhelmed. Irrigation by diesel pumps from shallow wells is then used to distribute the stored soil moisture; highly productive export crops of peppers, egg-plants, tomatoes and cucumbers are produced for distribution by sea and air (Plate 25).

The storage of floodwater underground is being developed increasingly in semi-arid countries. Current examples are from Saudi Arabia, where the Majmaah Dam is being built across the Wadi Namil, to the north of the capital city of Riyadh; and also on the Mediterranean coast the new Israeli port of Ashdod, built on the site of the ancient city, is developing recharge areas in the sand dunes to harvest the rain-season torrent flow of the Lakhish River.

Drainage of marshlands

The association of flat land and surplus water has in most countries offered to enterprising cultivators the opportunity to construct drains

194

and thus to reclaim marshes for their crops and pastures. The hydro-logical effects of such operations depend on the geological and topo-graphical circumstances which have formed the marsh. Contradictory examples are thus available for controversy. Are freshwater areas of wetland, swamp, marsh, vlei, slough, etc., a resource or an obstacle to development? Do they act as regulators of streamflow or are they obstructions, losing more in evaporation than they conserve by storage? The confusion arises because all of these terms are general descriptions of areas of similar ecological associations of water-loving plants, which may be sustained by entirely different physical causes of water surplus.

In steep mountain country under high rainfall, marshes are found behind natural rock barriers, forming natural reservoirs invaded by vegetation. The water levels are drawn down by evaporation in dry weather to provide some degree of storage for the regulation of spates in wet weather. Drainage by cutting through the rock barrier merely destroys the reservoir. Drainage by a partial reduction of the water level and irrigation by intercepting the stream as it enters the marsh, and leading it around the hillside to supply irrigation furrows to a central drain, can provide an agricultural resource, the disadvantage being that it remains vulnerable to flooding.

Where marshland is due to a loss of water velocity in a limited length of flat streambed the organic debris may be the only form of barrier, as in the papyrus swamps around Lake Victoria. Drainage of such marshes in the tropics has led to the rapid oxidation of the peat, leaving a coarse white sand of no agricultural interest. Skilful drainage, lowering the water table only a little, has been more successful, although a hazard encountered in Uganda was the presence of ferrous sulphide in the marshy soils. This gave no trouble while they remained wet, under reducing conditions, but as the soil dried oxidation produced sulphuric acid, causing the complete failure of crops on the drained land.

The reduction of any area of freely transpiring vegetation may be expected to reduce transpiration losses, and the hydrological effects of the draining of marshes are therefore usually an overall gain in water quantity with some loss of regularity of seasonal flow. For example, very large quantities of water are transpired by swamp vegetation in the Sudd area of the Southern Sudan, where the main channel of the Nile dis-perses into marshland. An engineer study of the 'Jonglei Canal' pro-posal has indicated that substantial increases in the flow of the Nile could be achieved by cutting a canal to drain these marshes and carry the river through them.

If, however, the marsh is formed by the outcropping of a water table,

195

as commonly occurs near the bases of hills, then the seepage from the springs already controls the flow rate, so that the effect of drainage is then to reduce transpiration without impairing the regular flow.

The mere removal of vegetation, to replace a swamp by an open-water surface, was at one time believed to reduce evaporation, but studies by Linacre and colleagues (1970) in Australia, and by Rijks (1969) in Uganda, as quoted in Chapter 3, have established by measurements of energy and of water-vapour flux rates that the shading of the water surface by a cover of reeds can result in a reduction in water use. From Chapter 3 it will be apparent that such results could be reversed on occasions when a dry warm wind passes over tall reeds wetted by rain. In most circumstances, however, there would be no hydrological gain in clearing swamp vegetation to replace it by an open-water surface.

Drainage effects on marshes subject to seasonal soil freezing, snow accumulation and snowmelt present complex problems of measurement and interpretation. These have been studied with great thoroughness in the USSR where vast areas of natural swamplands occur in the water-surplus climate of North-Eastern Europe. Professor K. E. Ivanov, of the State Hydrological Institute at Leningrad, has published sixteen years of very detailed water-balance observations on the wet, cold, sphagnum-moss swamps (Ivanov 1953, 1963). These swamps are being reclaimed by drainage, largely for timber production, and the scientific study has been thorough enough to give useful predictions of hydrological effects.

The swamps are convex lenses of peat and mosses, draining radially into peripheral channels. On the large rivers, swamps form on successive river terraces separated by banks of drained soil carrying pine forest (Fig. 29). The lenses of peat vary in thickness from a few centimetres near to the swamp edge to a maximum depth of some 10 m. The water tables lie near to the surface of the peat and follow the convex shape, draining towards the edges. Pine trees grow 15 to 20 m high on well-drained sites but are stunted in the swamps, attaining a height of 3 m or less. Ten Swamp Observation Stations are now measuring water balances, and the writer was able to visit the experiments at Zelenogorsk.

Here evapotranspiration is estimated from the measured radiation balance and the vapour-pressure gradient. The results are checked by floating monolith lysimeters of the Ouryvaev pattern, and by small weighed evaporimeters in replicate groups of eight. Instruments in the swamp are remotely recorded from a laboratory on higher ground and weirs are housed to prevent freezing. Field data for sixteen years over a wide variety of swamps have made it possible to estimate peak discharges from a combination of aerial photographs, simple field reconnaissance measurements, and the climatic characteristics of snowmelt obtained from the latitude and longitude. Maximum flows from snowmelt on these sphagnum-moss swamps are reduced by drainage, and the

Where the swamp occupies the whole of a divide

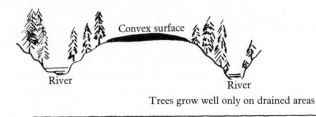

Trees grow well only on drained areas

Classical Siberian swamp profile on river terraces

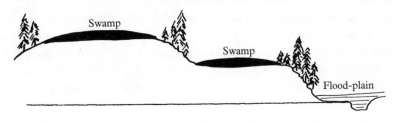

Swamp study at Zelenogorsk (on a terminal moraine formation)

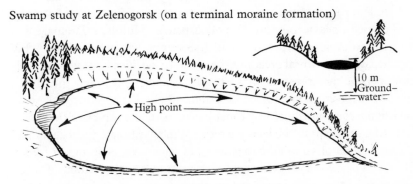

Fig. 29. **Studies of sphagnum-moss swamps in cold wet areas of North-Eastern USSR.**

Freezing and thawing of the swamps and of the snow-cover present complex problems for quantitative studies of drainage; the swamps are drained to improve the growth of pine trees. (Ivanov 1963.)

extent of the reduction is tabulated by Ivanov (1963) in terms of area, slope and peat characteristics of the swamp. In very dry years drainage can increase the maximum flow and, more important, the minimum flow is also increased. This effect becomes important in climates of water deficit, where swamp drainage improves overall water yield. Control of a high water table for irrigation of agricultural crops on the drained marshland was found, however, to result in increased water use and decreased average flow.

Effects of swamp drainage on water quality. This is another hydrological aspect of swamps in which generalisations are meaningless unless the geographical circumstances are specified. Where clean water trickles from springs into swamps, the water may become stained with decaying organic matter and, although the concentrations may be very slight indeed, both colour and taste may be affected. Except, perhaps, for the preparation of Scotch whisky, this has an adverse effect on water quality. In river systems, however, where solids are carried in suspension, the low velocity of water flow through swamps offers a valuable opportunity for the settling out of suspended solids. The filtration effect of the swamps on the Parana River system, mentioned in Chapter 2, provides a large-scale example. The drainage of swamps in such circumstances can have serious effects downstream. The drainage of the Huleh Swamps on the upper reaches of the River Jordan provides a substantial example. Here an area of about 60 km^2 was seasonally flooded and provided a settling basin for the runoff from some 1500 km^2 of headwater area (Neumann 1955). Much of the runoff is torrent flow from rugged country and carries a heavy concentration of suspended sediment. Drainage and irrigation development of the Huleh Swamps has produced a highly productive agricultural area, but the greater velocities in direct channels now permit the transport of suspended sediment through the Huleh area for deposition in the Sea of Galilee (Lake Kinnereth), with some additional soil contributed by canal banks. A recent survey of the inflow of fluvial sediment to the lake has resulted in an estimate of mean annual sedimentation of 30 million tons, 90% of which comes from the Jordan River. The lake is a very efficient sediment trap, and the volume of solids represents an annual layer 90 mm thick over the whole lake bottom (Schick 1968).

Effects of swamp drainage on flood control. Where swamps occur high in the watershed the effects of drainage of large areas can be a reduction in the delay of storm runoff, with a consequent liability to increase stormflow peaks downstream.

It is usually difficult to find the records with which to demonstrate such effects unequivocally, but a good example is from the long historical record of flood heights in the upper catchments of the rivers Wye and Severn, on the flanks of the Cambrian Mountains. Study of these records has confirmed the subjective impression that flooding in mid-Wales is increasing in frequency and seriousness (Howe, Slaymaker and Harding 1966). Records began in the eighteenth century and have been continuous since 1911. The authors showed that there had been, in both catchments, greater flood heights over the period 1940–64 than over the period 1911–40. During the earlier period a flood height of 5 m could be expected once in twenty-five years; during the later period this height was reached once in every four years. The basic reason has been a substantial increase in the number of rainfalls of over 2·5 inches per day (63 mm/day), which occur with twice the frequency in the second period as compared with the first. Changes of land use, by the drainage of peat swamps, both for peat-cutting and for afforestation, also appear to have had a direct effect. By plotting the flood peaks against drainage development, estimated as miles of channel per square mile of watershed, on logarithmic scales, the authors obtained a linear relationship. More evidence is expected from the flood studies now in progress at the Institute of Hydrology.

Ecological effects of marsh drainage. In countries of advanced technology drainage and development have progressively reduced the areas of wetland, destroying characteristically rich associations of marsh plants, wildfowl and other adapted species, so that some rare species are in danger of extinction. The need to preserve adequate areas of wetlands in Nature Reserves for biological study, and in National Parks for recreational observation, has in the past been difficult to convey to a public unfamiliar with the concept of preservation of the natural environment, but there has been a recent encouraging surge of interest on a world scale. A famous example, which has caused fierce controversy, is the drainage and diversion of flows from the Everglades in Florida. Lake Okeechobe is one of the major lake areas of the USA. It is very shallow and overflows frequently into the flat marsh country only a few metres above sea level. The peat of the Everglades forms nearly one million hectares, the largest known continuous area; it is from 0·5 m to 4 m deep. When drained it subsides at a rate of 25 mm per annum (Stephens 1956) and the dried peat can present a severe fire hazard in dry weather. The swamps have always varied in water level, but small pools and alligator holes then provided for survival of aquatic species. Drainage has led to the disappearance of alligators and much of the wetland flora and fauna from the areas affected. Preservation of a large area as a National Park is, however, now assured, although water levels will need careful control. A world check-list of 650 aquatic sites, with descriptions and reasons for their preservation, has recently been published for the International Biological Programme (Luther and Rzoska 1971).

10

Problems and priorities

The previous chapters have surveyed much encouraging technical progress towards the better joint development of land and water resources. This has been gained both by empirical experience and by scientific research and development. Countries at all stages of technical and social development are faced with the dilemma of balancing short-term advantages to standards of living against the safeguarding of the water supplies, soil fertility and countryside amenities of future generations. These problems are different in detail but equally acute and important for the most developed nations and for those at an early stage in the assault of modern commercial technology on traditional cultures.

In countries of advanced technology

The absolute priorities for the most developed countries of Europe, North America, Australia, Japan and elsewhere are to develop increased supplies of fresh water while reducing the pollution of existing resources. Implicit in these aims is the improvement of land use to secure better regulated flows of water free of soil. As already suggested in Chapter 2, these aims are consistent with increased countryside amenities for increasing populations with more leisure and greater mobility.

Watershed effects of urban and highway development. The roofs and pavements of a city represent the widest departure from the absorptive natural surface of vegetation and soil, so that the basic engineering design of a city must include drainage systems able to cope with very rapid accumulation of runoff water. As cities have grown, so the concentrations of surface stormwater have needed more accurate study, particularly by networks of sensitive recording raingauges. Studies in London by the Road Research Laboratory (Watkins 1962) since extended to tropical conditions in East Africa (Forsgate and Temiyabutra 1971) and others in Japan, especially in Tokyo, and in Washington (Wolman and Schick 1967) are typical of the increased attention the subject is receiving. The solutions are, in general, by direct application of known engineering principles and techniques, and the main problems are usually those of cost. An important matter of watershed planning, however, may be involved in disposing of the peak flows into natural stream channels. Large cities so outgrow their geography that their stormwater flows can greatly exceed the capacity of the natural channels; substantial flood-retarding storage basins are therefore needed.

Small towns and villages scattered throughout a large watershed have far less effect, e.g. in the one-million-hectare watershed of the River Thames above Teddington Weir there has been much recent development, including four entirely new towns created by regional planning. Comparison of surveys in 1939 and in 1962 showed an increase in impermeable surfaces of roads and buildings of some 6000 ha. This is, however, less than 0·5% of the agricultural area, so that it is not surprising that the long-term records of river flow, since 1883, showed no change resulting from this development (Andrews 1962).

Modern motorways are also adding very large collecting surfaces which can severely affect the flow regime of natural streams, but the effects can be calculated and provided for where the water authorities are alert to the dangers involved. Stabilisation of the vast earth embankments of modern motorway construction appears to be a neglected technique in Europe, and particularly in Britain, where temporary soil erosion is still tolerated to a surprising extent. Under the harsher continental climates of the USA, more elaborate, and doubtless more expensive, precautions are taken for the early establishment of a grass cover on exposed soil.

Industrial pollution. Attention was drawn in Chapter 1 to the critical problems of water pollution which now face all industrialised nations. As cities have outgrown their geographical facilities for both water supply and sewage disposal it has proved easier to supplement the input by piping water from a distance, than to solve the disposal problem for urban sewage and industrial effluent. Major rivers have thus been degraded to open sewers, even although they flow through the centres of wealthy cities, such as the Hudson through New York, the Thames through London, the Seine through Paris, and, most polluted of all, the Rhine through Cologne and the Netherlands. After more than a century of neglect of a steadily worsening problem, public attention in the more developed countries is concentrating increasingly on the correction of this misuse of watercourses. The costs, both for capital and recurrent expenditure, are so heavy as to be a major burden on modern civilisation. Throughout the USA, in spite of totally inadequate control of industrial effluent, the costs of disposal of public wastes already ranks third in municipal expenditure, being exceeded only by the costs of education and of highway construction (Rampacek 1968).

Where industrial effluents of toxic chemicals and heavy-metal sludges are excluded, the techniques of purifying sewage waters are well advanced. The purification is by gravity separation of solids and the chemical and biological breakdown of organic materials under anaerobic conditions in settling tanks, followed by the completion of the biological oxidation of organic matter in well-aerated ponds, streams and rivers, with final filtration and sterilisation, usually by chlorination, in water treatment plants. The final stages of aeration, carried out in large lakes or lagoons, have been developed in California by the Santee City authorities into public amenities for sailing and fishing. In a similar climate the city of Tel Aviv, under Israel's dominant problem of water scarcity, processes the whole of the municipal sewage in lagoons among the sandhills, finally spreading the water over infiltration areas to be stored underground. By pumping from a distant point in the aquifer both dilution and prolonged storage serve to attenuate the bacterial populations to levels which can be safely eliminated by the city's water treatment plant. Under the favourable climates of California and of Israel the area of oxidation ponds is about four acres ($1 \cdot 6$ ha) per 10,000 population.

Great Britain was early in the field of legislation on pollution, following a Royal Commission in 1865. The reluctance of both public authorities and private enterprise to spend large sums on the purification of industrial wastes has unfortunately over-ridden the good intentions of the legislation. The Water Pollution Laboratory, established in 1927, has greatly improved the formal knowledge of the situation and the techniques to remedy it. Nevertheless, a government survey in 1958 of all rivers whose flows exceeded 4500 m^3/day showed 6% of the total river length to be grossly polluted, 21% needing action for protection and improvement, and 73% unpolluted or recovering (McNaughton 1962).

A very heartening example of progress in this formidable task comes from London, where the Thames, in the decade following the Second World War, reached a severely anaerobic condition amounting to a public nuisance. Research findings by the Water Pollution Laboratory were taken up by the London County Council, with an expenditure of over $100 million in ten years. Recovery has been steadily achieved and by 1967, for the first time, not a single water sample was found to be devoid of oxygen at any of the twenty-nine weekly-sampled sites from 10 miles upstream to 50 miles downstream of London Bridge. Fig. 30 illustrates the progress to 1969. Many species of fish have returned (Plate 26). A survey in 1958–9 showed no living fish in the 35 miles

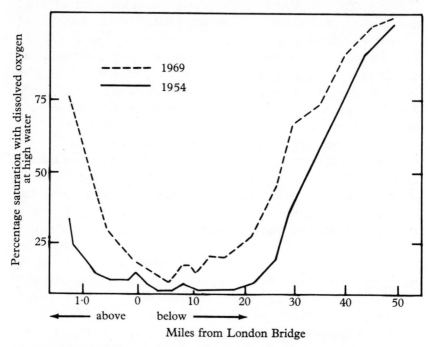

Fig. 30. **Biological recovery of the river Thames as a result of sewage control** (% **saturation with dissolved oxygen**).

In 1930–59 the discharge of both sewage and industrial wastes into the Thames deprived the water of oxygen. Fish were eliminated. Ten years of costly improvements in water purification were needed before discharge into the river gave the encouraging progress shown in 1969. Fish have returned (see Plate 26). (From First Report of Royal Commission on Environmental Pollution 1971.)

downstream from Richmond but in 1968–9 catches of forty-two species were reported by the British Museum of Natural History.

Industrial wastes of ever-increasing chemical complexity present a far more intractable problem than domestic sewage. In particular the salts of heavy metals, copper, lead, zinc, iron, chromium and cadmium, produce highly toxic sludges which are difficult to dispose of even when trapped and concentrated. It is important that toxic wastes, if banned from rivers, should not be ponded when they can infiltrate into groundwater. A clearly-stated summary of Britain's worst problem, the rivers Tame and Trent which drain Birmingham and the industrial concentrations of the Midlands, is given by Tinker (1971) as a report on a symposium of the Institute of Water Pollution Control. A more detailed discussion by the Royal Society has been edited by Russell and Gilson (1972).

In the USA, in spite of declarations by successive Presidents in 'State of the Nation' addresses that national efforts were to be made to

Plate 26. Success in the control of pollution: fish return to the river Thames.

A direct and encouraging result of the improvements in the water quality reported by the Royal Commission on Pollution (see Fig. 30). A sea trout of $3\frac{3}{4}$ lb caught above Woolwich in February 1971. (British Museum of Natural History.)

reduce river pollution, and in spite of the expenditure of some $250 million in the past five years on research into the control of water pollution, no substantial positive achievements in the cleansing of rivers appear to have yet been claimed on the basis of published data. Even in New York, waste discharges contribute 50% of drought flow of the Hudson River south of the city line and one-third of the sewage runs untreated into the river (Howells 1972). On the Rhine, a major incident in which a very large number of fish were killed by industrial effluent during 1970 has caused deep concern and the renewal of efforts to secure agreement on control of pollution in this international river.

A physical form of pollution, by waste heat from power stations, has caused some concern in the past two decades as power consumption has risen in many industrial countries. The alternatives are either to run the heated water back into the river or to evaporate water from cooling towers. In the former case the rise in temperature is reported by Langford and Aston (1972) to have less undesirable biological effects than had been expected, and the water is rich in oxygen. In the latter case the water losses are substantial, and since pollutants are not evaporated, they are returned to the river in concentrated form. There are, however, in a few situations, encouraging possibilities of using waste heat to maintain fishponds or even shallow sea inlets at an optimum temperature for fish production.

Pollution from modern agriculture. Of the heavy dressings of fertilisers applied to both cropland and pasture in modern farming systems, the nitrogen is very soluble and is leached in large quantities into the drainage water. A recent study, by the US Agricultural Research Service and the University of California, of a watershed of 384 ha (960 acres) under irrigated citrus orchards, reported drainage losses of 45% of the nitrogen from fertilisers. Concentrations of up to 87 ppm of nitrogen were measured in the drainage water. A survey of the irrigation lands in the Riverside region showed similar losses from other watersheds (Bingham, Davis and Shade 1971). Phosphates are mainly retained by the soil. In measurements of the nutrient contents of waters in reservoir storage, the Water Pollution Research Laboratory of the UK estimated that five-sixths of the nitrogen comes from agricultural land; of the phosphorus about half comes from detergents and most of the rest from industrial wastes (Downing 1965). This enrichment of the plant nutrient contents, termed 'eutrophication', supports excessive growths of algae which cause difficulties in water purification.

Cattle manure was once a key factor in maintaining soil fertility for 'high farming' but labour costs of manuring are now excessive: the ready supply of fertilisers and their cheaper application by mechanical spreaders have greatly reduced the demand for dung. Concentrations of livestock for fattening at feedlots and the even more intensive penfeeding of 'factory farming' methods now produce large quantities of animal wastes from which the drainage is a pollution hazard both to surface and to groundwaters. In the UK alone some 12 million cattle, 7 million pigs and 127 million poultry spend all or part of their lives indoors (Ashby 1971). Tomlinson (1971) points out that the nutrients in animal excreta in the UK amount annually to eight to ten times the total output from human sewage and domestic wastes (Table 7).

Table 7. Nutrients in animal excreta and in sewage in the UK.

The concentration of livestock in modern methods of animal production is leading to disposal problems which are formidable in proportions, as seen by comparison with domestic sewage.

	Total nutrients (tons per annum)		
	N	P	K
Animal excreta	800,000	150,000	1,000,000
Human sewage and wastes	102,000	18,000	21,000

The problem of keeping these nutrients out of the drainage water and of returning them to the land by economical methods is serious and in the UK the Agricultural Research Council is supporting research on treatment of animal wastes and the agricultural use of the nutrients. Thus in rural as well as in urban areas modern industrial techniques need modification to prevent water pollution. In the UK, a Royal Commission on the Pollution of the Environment has recently completed the first stage of a study under the distinguished chairmanship of Sir Eric Ashby (1971). Encouraging data are quoted on the reduction of smoke and of chemical air pollution in cities, and the improvement in the quality of water in the River Thames is described. However, the report stresses that there is a great deal more to be done. 'Not only is the state of some of our rivers depressing: too many of them are so polluted that they cannot be used to meet our growing needs for water.' The Commission concluded that it would need to keep under review action to improve Britain's rivers, and regarded three issues as particularly important:

'(i) the integration, under a single authority in each river region, of the administration of rivers and sewage treatment;

(ii) the improvement of qualifications and training of those who control water pollution;

(iii) the application of up-to-date process and chemical engineering to the design and operation of sewerage and sewage plant.'

The political aspect of the problem is well summarised by the Royal Commission in a single sentence:

> 'Pollution's main economic characteristic is that its costs are not usually borne by the polluters, so that production is often pushed beyond the socially optimum point; there is inadequate incentive to allot sufficient resources to reducing pollution; and certain producers and consumers benefit at the expense of the victims of the pollution.'

Watershed control and countryside amenity. Water supply is already recognised as a dominant factor in planning the land use of highly populated and industrial countries. Reservoirs already cover significant areas of valley bottomlands of high agricultural potential, but more land is released by the continuous increase in the efficiency of agriculture, which concentrates production on a steadily diminishing area. The differences between the water use of agricultural lands and forest plantations are already important, and from preceding chapters it is clear that useful estimates can already be made of such differences. Even in the mild climate of Britain local conditions decide the issue and forest plantations in the very wet steep mountain areas of Wales may be beneficial, with the advantages of streamflow regulation outweighing the differences in water use, whereas in drier and flatter country in the south-east of England the extra water use may be a serious disadvantage.

The need to provide amenities for large numbers of urban visitors is now accepted in country planning; public access to both watershed forests and reservoirs is now not only accepted but sometimes invited as a source of revenue. In the TVA system in the USA, for instance, the writer was greatly impressed by the way in which construction workers for major reservoir projects were housed in well-built camps which then remained for holiday-makers who fished, sailed and swam in the lakes created. The efficiency of modern water-treatment plant is now adequate to safeguard water supplies under such popular visitation. Pine plantations also, in addition to absorbing rainfall and controlling streamflow, can often absorb far more holiday-makers in some degree of peace and solitude than would be able to enjoy themselves on the same area of open moorland.

In developing countries

Urban development and pollution. Developing countries are faced with equally acute difficulties of protecting their natural resources of land and water, but under greater pressures of population increase and with a lesser degree of social organisation for the protection of public interests. Problems of water pollution are acute in cities surrounded by ever growing populations in camps with rudimentary sanitation; those of control of industrial pollution, difficult enough in technically developed countries, are exaggerated when new industries grow up in the absence of control legislation and technical inspectorate. These are scarcely matters needing research. Indeed the technical solutions are often known to the governments concerned, and technical advisers are readily available from sources of international aid. The problems are of government and municipal organisation for the application of known solutions, and the political task of persuading an unheeding population of their importance.

Highway drainage. Administrators of technical aid can, however, do much to persuade political leaders of the importance of conserving their country's irreplaceable natural assets. A good example is in the loan or grant of funds for the construction of highways. By engineering tradition the highway designer leads his road drains to natural creeks or gullies when possible, but otherwise discharges them down slopes, along drains prepared for a few yards only, i.e. enough to protect his road from erosion damage. Some of the most acute soil erosion to be seen in Africa starts from the out-turns of new highways built with modern equipment. The funds should be sufficient to improve the drainage ways to take the increased flows without severe soil damage.

Similarly, international aid for irrigation developments should include the control of headwater catchment areas for the reasons set out in Chapter 2, with financial provisions for the necessary patrolling of critical watersheds.

Control of land use in streamsource watersheds. For the first two decades after the last World War the dominant objective of international aid in the natural resources field was to mobilise scientific and technical missions in order to bring modern knowledge to solve the urgent problems of the developing countries. For the past decade there has been an increasingly baffled tone in the resulting technical reports, e.g. FAO Meeting Reports (1958–9) for the Middle East ranged over Iran, Iraq, Syria, Lebanon, Jordan, Egypt, Sudan, Libya and Tunisia, and came to the general conclusion that 'specialists in the region already know what is technically possible, but that great difficulties, which cannot presently be surmounted, are of an administrative, political and social nature'. Similarly Kaul and Thalen (1971), reporting from Iraq, describe the need for grazing control to prevent the destruction of the sparse vegetation of semi-desert lands by the rising population of people and livestock, and conclude that 'it is difficult to apply in practice because of socio-economic conditions'. Ceylon is a good example of a tropical country with almost every advantage for good land-use management. There are well-developed technical services based on an advanced educational system and good contacts with overseas science. The advantages of a highly-developed cash-crop industry with major export earnings are backed by great reserves of uninhabited land amounting to some two-thirds of the island, with good overall water resources capable of development.

Yet in spite of all these advantages, Ceylon is caught up in a major struggle to control the land-use problems created by its violent population expansion. The population doubled between 1931 and 1963, and will have doubled again before the end of the century. The majority of the increasing population lives by subsistence agriculture. In spite of government schemes to open up new districts by organised settlements, the population is overspilling into Crown Land Forest and Forest Reserves, stripping off the protective vegetation without any form of soil protection. Creation of new economic opportunity without organisation of land use has accelerated this process, as in the sudden expansion of tobacco growing on land too steep for economic tea production. This has created accelerated soil erosion damage over some 3000 acres of slopes exceeding 100% (45°), on which not even contour planting is yet practised. The immediate effect is damage to the existing capital development in irrigation schemes below the eroded areas. There is an urgent need to expand food-crop production, since the island has ample land, water and labour to grow the food which is now imported. At present, however, much of the expanding food production is on steep slopes without protection, in high-rainfall streamsource areas which supply reservoirs for irrigation. The increased sedimentation of these reservoirs is reducing their capacity to supply their dependent irrigation areas and is thus causing their earlier drying-up. Soil-conservation practices are not yet supported by any government subsidies. Such subsidies have been found necessary in most countries which operate successful soil-conservation programmes. They are probably essential to safeguard the future of Ceylon's land and water resources.

The land-use problems of many developing countries are intensified by maldistribution of population, but centuries of tribal warfare in areas of nomadic grazing or precarious cultivation have fixed boundaries which are still fiercely defended; administrators cannot therefore re-settle excess populations from overcrowded country into more sparsely occupied land. In discussion with a provincial governor in Ethiopia the writer pointed out that a very steep hillside had been newly opened to cultivation and was already eroding. He gravely agreed and said that he had advised against use of these steep slopes. Since he evidently carried real authority he was asked why his advice was not heeded. 'Where can I tell them to go?' he replied. The more suitable land lay beyond a traditional boundary. Recalling the fierce resistance when new towns are created by planners in technically developed countries, one could only agree that this was not a problem which science could solve.

International aid. We can, in the meantime, greatly improve the efficiency of our technical aid by recognising the watershed as the logical unit within which to co-ordinate all the separate efforts by which different technical specialists seek to offer help. In this way the highly popular items, such as medicine and education, can be made part of a 'package deal' with soil conservation, the protection of vegetation against overgrazing and excessive burning, crop sanitation and all the necessary disciplines of a higher agricultural productivity. The indi-vidual cultivator must be rewarded economically for such discipline if it is to reach stability, and a vital part of overseas aid should go to make such rewards possible.

This will often require the improvement of communications and of transport, the provision of storage for fertilisers and for harvests, the improvement of marketing and the provision of credit. Much overseas technical aid has little effect because projects aimed to remove each separate limitation on agricultural production are operated inde-pendently, by different agencies, and applied in different watersheds. The organisational problems are formidable, but some countries are taking experimental steps towards such integrated forms of bilateral aid. The principle is not yet current among donor nations and is becoming accepted, only very slowly, by United Nations agencies, which are organised under separately specialised technologies. The Development Projects financed by the United Nations Development Programme through the FAO already provide the framework for such multi-disciplinary efforts to improve complete watersheds.

Economic development. The role of science and technology in the battle against poverty in the less developed countries has been thoroughly examined by Graham Jones (1971) in a study sponsored by the International Council of Scientific Unions and supported by an international committee under the chairmanship of Lord Blackett, then President of the Royal Society of London.

The conclusions of this wide-ranging study apply very specifically to the development and protection of land and water resources. Thus, in both agriculture and industry, improvement in the practical use of existing knowledge by strengthening the supporting services and by raising the general level of productive competence is likely to be more important than specialised research. 'Management is probably the most important factor.' The need to apply improved management to watersheds and to the safeguarding of water supplies from pollution is an urgent example of the application of Graham Jones's conclusions.

The spread of subsistence agriculture. In the developing countries in the tropical climates where, as the foregoing chapters have shown, the penalties of land misuse strike earlier and harder, there is little sign of man's winning this grim race against time. Ultimately the hydrological behaviour of watersheds must reflect the pressure of the population upon them. The damage to soil and water comes from destruction of protective vegetation by the spread of unorganised and unskilled subsistence agriculture in ever steeper and drier country (Fig. 31). The effects, so well described by Plato 2300 years ago (cf. Chapter 2) are gathering momentum. Already in South-East Asia and in parts of Latin America the rates of human increase have outrun both the powers of their governments and the help of friendly nations, to maintain their standards of living, which have deteriorated sadly. Already half of the total world population is underfed. Addressing the UN Biosphere Conference in Paris in 1968, the Secretary-General of FAO, Dr Boerma, warned that although agriculture could feed mankind at present population levels, no effective progress towards improving living standards would be possible if populations continued to rise at present rates: while population keeps pace with agricultural increases there will be no improvement of subsistence agriculture (Boerma 1968). By the year 2000 there will be six additional members of the underdeveloped areas of the world for every additional member of the developed countries (US National Academy of Sciences 1969).

The increasing pressure of subsistence population on lands already too overcrowded for efficient agriculture is emphasised by the FAO Indicative World Plan; this shows that even in the next fifteen years 85% of the children born throughout the world will be in the less developed

212

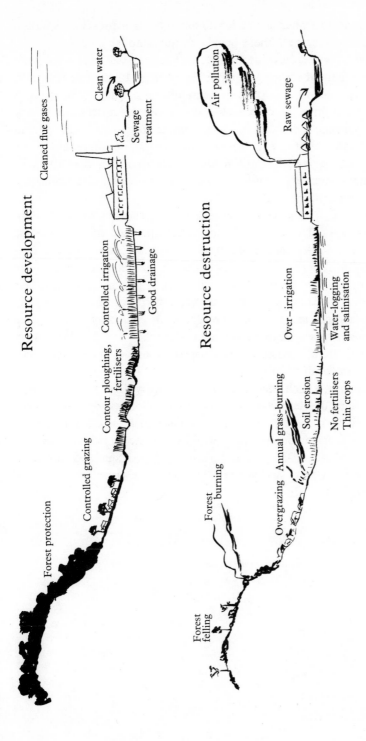

Resource development

Forest protection

Controlled grazing

Contour ploughing, fertilisers

Controlled irrigation

Good drainage

Cleaned flue gases

Clean water

Sewage treatment

Resource destruction

Forest felling

Forest burning

Overgrazing

Annual grass-burning

Soil erosion

No fertilisers
Thin crops

Over-irrigation

Water-logging and salinisation

Air pollution

Raw sewage

Fig. 31. (a) **Resource development.** (b) **Resource destruction.**

Both processes are at work and are readily recognisable. The scientific knowledge to change from destruction to development is already available. Political leadership is probably more important than technical aid in progressing from (b) to (a).

countries, most of them dependent upon subsistence-level agriculture. This will represent 50% more people on the same land.

The most widespread destructive influence on the hydrological balance of mankind's water supplies is the struggle by increasing human populations to survive by primitive agricultural methods (Fig. 31). All mankind must be alerted to realise that if we are to remain in any tolerable balance with our soil and water resources then human numbers cannot continue the present mindless rate of increase.

It may well prove to be the medical scientist rather than the soil-conservation engineer whose progress will eventually determine the condition of our watersheds.

Of two very experienced observers of the problems of subsistence agriculture René Dumont (1966) describes the outcome as 'a tragic economic and food situation and thus an explosive political situation', while William Allan (1967) concludes that 'the outcome remains in the balance; increasing experience and efficiency of governments and their technical services, limitation of population growth, continued economic and technical aid *and above all the popular acceptance of the discipline of development* may lead to a fundamental change'.

Those who are concerned with the natural resources of land and water can take heart from the progress that has already been made in preventing the damage, which Plato described so clearly, from bringing about the predictions of Malthus, but we must recognise that the battle is not yet won. Even if we can assume a stabilisation of world population, much more efficient organisation of aid from the wealthier countries and more effective self-help from the less developed countries are necessary to ensure that the land and water resources of our planet will continue to support us all.

Recommended reading

The 385 references to the scientific literature are the sources of evidence for the arguments advanced, and will give further information on the points under discussion. The following are suggested for detailed reading for closer acquaintanceship with both the problems and the science of land and water resources.

Developing countries:

Allan: *The African Husbandman* (1965). Oliver and Boyd, London.
Dumont: *African Agricultural Development* (1966). UN Pub. E/CN, 14/342. New York.
US National Academy of Sciences: *Resources and Man* (1969). Pub. No. 1703. Freeman and Co., San Francisco.
Jones: *The Role of Science and Technology in Developing Countries* (1971). Oxford University Press.

Agriculture:

Webster and Wilson: *Agriculture in the Tropics* (1966). Longman, London.
Russell: *Soil Conditions and Plant Growth* (11th Edition, 1973). Longman, London.
US Dept. of Agriculture: Yearbooks. Washington, DC.

Irrigation:

Reifenberg: *The Struggle Between the Desert and the Sown* (1955). Hebrew University, Jerusalem.
Hagan, Haise and Edminster (Editors): *The Irrigation of Agricultural Lands* (1967). American Society of Agronomy. Madison, Wisconsin.
Clark: *The Economics of Irrigation* (1967). Pergamon Press, Oxford.

Water engineering:

Valentine: *Water in the Service of Man* (1967). Penguin Books, Harmondsworth.

Hydrological research:

Penman: *Vegetation and Hydrology* (1963). Commonwealth Bureau of Soils, Harpenden.

Sopper and Lull (Editors). *Forest Hydrology* (1967). Pergamon Press, Oxford.

Broughton: *Effects of Land Management on Quantity and Quality of Available Water* (1970). Australian Water Research Council. Res. Project 68/2.

Reifsnyder and Lull: *Radiant Energy in Relation to Forests* (1965). USDA Forest Service, Washington, DC.

References to literature

Ackerman, W. C. (1955). Influences of reafforestation and erosion control upon the hydrology of Pine-tree Branch Watershed 1941–50. TVA Tech. Monog. No. 86. 95 pp.

A'Hafeez, A. T. and Hudson, J. P. (1965). A cheap weighable container for measuring evaporation in the Sudan. *Exp. Agric.* **1**, 99–106.

Albert, F. A. and Spector, H. A. (1955). The muddy Chattahoochee. *USDA Yearbook*, p. 205. Washington, DC.

Alden, E. F. and Brown, F. J., Jr (1965). A prefabricated flume for gauging ephemeral streams. US For. Serv. Res. Note RM55. USDA.

Allan, William (1965). *The African Husbandman.* Oliver and Boyd, London.

Amramy, A. (1967). Re-use of municipal waste water. *Proc. Int. Conf. Water for Peace* **2**, 421. Washington, DC.

Andrews, F. M. (1962). Some aspects of the hydrology of the Thames Basin. *Proc. Inst. Civ. Engrs* **21**, 55–90 and **24**, 247–287.

Ångström, A. (1925). The albedo of various surfaces of ground. *Geografiska Annaler*, H4, 323–342.

Arkley, E. J. (1964). Relationships between plant growth and transpiration. *Hilgardia* **34**, 559–583.

Ashby, Sir Eric (Chairman) (1971). First Report of the Royal Commission on Environmental Pollution. HMSO Cmnd 4585.

Australian Bureau of Meteorology. *Australian Meteorological Observers Handbook.* (1961). *Evaporation in Australia,* Hounan, C. E.

Bailey, R. W. and Copeland, O. L. (1960). Low flow discharges and plant cover relations in two mountain watersheds in Utah. Pub. No. 51, IASH Comm. of Surface Waters, 267–278.

– , Craddock, G. W. and Croft, A. R. (1947). Watershed management for summer flood control in Utah. USDA Misc. Pub. 639.

Baker, S. and Dill, H. W. (1971). USDA Econ. Res. Handbks 384, 409.

Barry, R. G. and Chambers, Ruth E. (1966). A preliminary map of summer albedo over England and Wales. *Quart. J. Roy. Met. Soc.* **92**, 543–548.

Bates, C. G. and Henry, A. J. (1928). Forest and streamflow experiment at Wagon Wheel Gap in Colorado. *US Monthly Weather Review* Supp. No. 36.

Bawden, M. G. (1967). Applications of aerial photography in land system mapping. *Photogram. Record* 5, 461–464, 469–473.

Bell, J. P. and McCulloch, J. S. G. (1966). Soil moisture estimation by the neutron scattering method in Britain. *J. Hydrol.* 4, 254–263.

(1968). Soil moisture estimation by the neutron method in Britain – a further report. *J. Hydrol.* 7, 415–433.

Bellani, A. (1836). Il colletore del calorico. *Annali delle Scienze del Regno Lombardo-Veneto*, p. 200.

Bennett, H. H. (1939). *Soil Conservation.* McGraw-Hill, NY and Lond.

Benton, G. S., Blackburn, R. T. and Snead, V. O. (1950). The role of the atmosphere in the hydrological cycle. *Trans. Amer. Geophys. Un.* 31, 61–73.

Bernard, A. E. (1945). *Le Climat Ecologique de la Cuvette Centrale Congolaise.* INEAC, Brussels.

(1953). L'Evapotranspiration annuelle de la forêt equatoriale congo-laise et son influence sur la pluviosité. *Comptes Rendus*, IUFRO Congress, Rome, 201–204.

Berwick, P. D. and Sumner, D. J. (1968). An accurate hydraulic-pneumatic weighing lysimeter for general field use. *Agric. Met.* 5, 5–16.

Bettany, E., Blackmore, A. V. and Hingston, F. J. (1964). Aspects of the hydrological cycle and related salinity in the Belka Valley, Western Australia. *Aust. J. Soil Research* 2, 187–210.

Bickmore, D. P. (1972). Experimental maps. Proc. Comm. Survey Offr. Conf.

Bingham, F. T., Davis, S. and Shade, E. (1971). Water relations, salt balance and nitrate leaching losses of a 900 acre citrus watershed. *Soil Sci.* 112, 410–418.

Birch, H. F. (1960). Soil drying and soil fertility. *Trop. Agric.* 37, 3.

Blackie, J. R. (1972). Hydrological effects of a change in land use from rain forest to tea plantation in Kenya. *Symp. Rep. and Exp. Basins.* Wellington, N.Z., *Bull. Int. Ass. Sci. Hydrol.* (in press).

– and Rawlings, M. H. (1972). Ann. Rept, Inst. Hydrol. UK.

Blaney, H. F. and Criddle, W. D. (1950). Determining water require-ments in irrigated areas from climatological and irrigation data. USDA, Soil Cons. Serv. Tech. Paper 96.

Blore, T. W. D. (1966). Further studies of water use by irrigated and unirrigated Arabica coffee in Kenya. *J. Agric. Sci.* 67, 145–154.

Bochkov, A. F. (1963). Influence of agrotechnical measures and protective forestry on river flow in forest-steppe and steppe zones. Trans. No. 127, State Hydrol. Inst., Leningrad.

Boerma, A. H. (1968). Food requirements and production possibilities. Final Report, Biosphere Conference. UNESCO, Paris.

Bouyoucos, G. J. and Mick, A. H. (1940). An electrical resistance method for the continuous measurement of soil moisture in field conditions. *Michigan Agric. Exp. Sta. Tech. Bull.* 172.

Bowen, E. G. (1966). The effect of persistence in cloud-seeding experiments. *J. Appl. Met.* **5**, 156–159.

Bowen, H. C. (1962). Discussion on hydrology of the Thames Basin. *Proc. Inst. Civ. Engrs*, **21**, 250.

Branson, F. A. (1956). Range forage production changes on a water spreader in SE Montana. *J. Range Management* **9**, 187–191.

Briggs, L. J. and Shantz, H. L. (1914). Relative water requirement of plants. *J. Agric. Res.* **3** No. 1, 63–87. USDA.

(1916). Hourly transpiration rate on clear days as determined by cyclic environmental factors. *J. Agric. Res.* **5**, 583–649.

Brohier, R. L. (1934). *Ancient Irrigation Works in Ceylon.* 3 Vols., Govt Press, Colombo.

Brunt, M. (1961). Air photography in land reclamation. *World Crops* **13**, 175–178.

Budyko, M. I. (1958). The heat balance of the Earth's surface. Translated from the Russian by Nina Stepanova. US Bureau of Commerce, Washington, DC.

Burger, H. (1943). Einfluss des Waldes auf den Stand der Gewasser. *Mitt. Schweitz. Anst. forst, Versuch sw.* **23**, 167–222.

Burvill, G. H. (1956). Salt Land Survey 1955. *J. Agric. W. Australia* **5**, 113–119.

Buttlar, H. V. and Went, I. (1958). Groundwater studies in New Mexico using tritium as a tracer. *Trans. Amer. Geophys. Un.* **39**, 660–668.

California Water Board (1957). *The California Water Plan.* Bull. No. 3.

Cashmore, W. H. and Hawkins, J. G. (1957). Tillage equipment and soil conservation. *J. Inst. Agric. Eng.* **13**, 20.

Childs, E. C. (1969). *An Introduction to the Physical Basis of Soil Water Phenomena.* John Wiley and Sons, London. 493 pp.

Chung-Ming Wong (1972). Report of US Office of Saline Water for 1971.

Cliff, E. P. (1958). The care and use of national forests. *USDA Yearbook*, 392–401. Washington, DC.

Cole, J. A. (1969). Hydrology and water resource development. Tech. Memo. 49, Water Research Association, Marlow, Bucks.

Costin, A. B. (1967). Management opportunities in Australian high mountain catchments. *Proc. Int. Symp. For. Hydrol.*, 565–577. Penn. State University (Pergamon Press).

– , Gay, L. W., Wimbush, D. J. and Kerr, D. (1961). Studies in catchment hydrology in the Australian Alps. III. Preliminary snow investigations. CSIRO. Div. Plant Ind. Tech. Paper No. 15.

– and Wimbush, D. J. (1961). Studies in catchment hydrology in the Australian Alps. IV. Interception by trees of rain, cloud or fog. CSIRO Div. Plant Ind. Tech. Paper No. 16.

Cotta, R. D. (1963). Influencia sobre el Rio Parana del material solido transportado por el Rio Bermejo. Com. Nacional d. Rio Bermejo, Buenos Aires, Pub. No. 92 E.H.

Courvoisier, P. and Wierzejewski, H. (1954). Das Kugelpyranometer Bellani. *Arch. Met. Geophys. Biol. B: Allgemeine und biologische Klimatologie* 5, 113–114.

Croft, A. R. and Bailey, R. W. (1964). Mountain water. Pub. US Forest Service, Ogden, Utah.

– and Ellison, L. (1960). Watershed conditions on Big Game Ridge. Station Report, US Forest Service, Ogden, Utah.

Culler, R. C. (1961). Hydrology of stock-water reservoirs in the Upper Cheyenne River Basin. Geol. Survey Water Supply. Paper 1. 531A. US Govt Printing Office, Washington, DC.

Cummings, N. W. (1940). The evaporation-energy equations and their practical application. *Trans. Amer. Geophys. Un.* 21, 512–522.

Currey, D. T., Webster, A. and Macleod, D. J. (1967). Salinity problems of the Murray Irrigation Areas. State Rivers and Water Supply Commission, Victoria, Australia.

Dagg, M. (1970). A study of the water use of tea in East Africa using a hydraulic lysimeter. *Agric. Met.* 7, 303–320.

– and Blackie, J. R. (1965). Studies of the effects of changes in land use on the hydrological cycle in East Africa by means of experimental catchment areas. Bull. IASH, X^e année, No. 4, 63–75.

(1970). Estimates of evaporation in East Africa in relation to climatological classification. *Geog. J.* 136, 227–234.

– , Hosegood, P. H. and McQueen, M. (1967). Rooting habits of East African grasses. *EAAFRO Record of Research*, Nairobi, 18–24.

– and Macartney, J. C. (1968). The agronomic efficiency of the NIAE tied ridge system of cultivation. *Exp. Agric.* 4, 279–294.

– and Pratt, M. A. C. (1962). Relation of stormflow to incident rainfall. *E. Afr. Agric. and For. J.* **27** (Special Issue), 31–35.

– , Woodhead, T. and Rijks, D. A. (1970). Evaporation in East Africa. *Int. Assoc. Sci. Hydrol. Bull.* **15**, No. 1, 61–68.

Dalton, J. (1802). Experimental essays on the constitution of mixed gases. *Mem. and Proc. Manchester Lit. and Phil. Soc.* **5**, 535–602.

Decker, J. P., Gaylor, W. G. and Cole, F. D. (1962). Measuring transpiration of undisturbed tamarisk shrubs. *Plant Physiol.* **37**, 393–397.

– and Wetsel, B. F. (1957). A method for measuring transpiration of intact plants under controlled light, humidity and temperature. *Forest Sci.* **3**, 350–354.

Dortignac, E. J. (1967). Forest water yield management opportunities. *Proc. Int. Symp. For. Hydrol.*, 579–592. Penn. State University (Pergamon Press).

Downes, R. G. (1961). *Conservation in the Eppalock Catchment*. SCA, Kew, Victoria, Australia.

Downing, A. L. (1965). Water pollution and the municipal engineer. *J. Inst. Municip. Eng.* **92**, 185.

Duley, F. L. (1939). Surface factors affecting the role of intake of water by soils. *Proc. Soil Sci. Soc. Amer.* **4**, 60.

Dumont, René (1966). *African Agricultural Development*. UN Pub. E/CN, 14/342. New York.

Dunin, F. X. and Downes, R. G. (1962). The effects of subterranean clover and Wimmera ryegrass in controlling surface runoff from four-acre catchments near Bacchus Marsh, Victoria. *Austr. J. Exp. Agric. and An. Husb.* **2**, 148–152.

Dyer, A. J. (1961). Measurements of evaporation and heat transfer in the lower atmosphere by an automatic eddy-correlation technique. *Quart. J. Roy. Met. Soc.* **87**, 401–412.

– and Maher, F. J. (1965). Automatic eddy-flux measurements with the Evapotron. *J. Appl. Met.* **4**, 622–625.

Ekern, P. C. (1964). Direct interception of cloud water at Lanaihale, Hawaii. *Proc. Soil Sci. Soc. Amer.* **28**, 419–421.

England, C. B. and Coates, M. J. (1971). Component testing within a comprehensive watershed model. *Water Resources Bull.* **7**, 420–427.

Engler, A. (1919). *Mitt. d. Eidg. Anst. f. d. forstliche Versuchwesen.* Bd XII.

Evanari, M., Shanan, L. and Tadmore, N. H. (1968). Runoff farming in the desert (I and II). *Agron. J.* **60**, 29–32 and 33–38.

FAO Meeting Reports (1958–9). Meetings on grazing and fodder resources of the Near East. FAO, Rome.

FAO (1965). Soil erosion by water; some measures for its control on cultivated lands. Agric. Dev. Papers No. 81. FAO, Rome.

Farbrother, H. G. (1960). Developments in farm mechanisation at Namulonge. *Emp. Cott. Grow. Rev.* **37**, 274.

Federer, C. A. (1965). Sustained winter streamflow from ground melt. US For. Serv. Res. Note NE, 41.

– and Tanner, C. B. (1965). A simple integrating pyranometer for measuring daily solar radiation. *J. Geophys. Res.* **70**, 2301–2306.

Federov, S. F. (1957). Evaporation from forests. *Trudy Gosudarstvennogo gidrologicheskogo instituta*, No. 59.

Fitzpatrick, E. A. (1968). An appraisal of advectional contributions to observed evaporation in Australia using an empirical approximation of Penman's potential evaporation. *J. Hydrol.* **6**, 69–94.

Forsgate, J. A., Hosegood, P. H. and McCulloch, J. S. G. (1965). Design and installation of semi-enclosed hydraulic lysimeters. *Agric. Met.* **2**, 43–52.

– and Temiyabutra, S. (1971). Rainfall and runoff from an industrial area in Nairobi, Kenya. Road Research Lab. Report LR 408. Dept Environment, Crowthorne, Berks.

Fourcade, H. G. (1942). Effects of the incidence of rain on the distribution of rainfall on unlevel ground. *Trans. Roy. Soc. S. Afr.* **29** (3), Cape Town.

Francois, T. (1953). Grazing and forest economy. *For. and For. Prods. Studies* No. 4. FAO, Rome.

Frenchman, M. (1971). The desert yields a rich food crop at a price. *The Times*, London, 21 Dec. 1971.

Fritschen, L. J. (1963). Construction and evaluation of a miniature net radiometer. *J. Appl. Met.* **2**, 165–172.

Funk, J. P. (1959). Improved polythene-shielded net radiometer. *J. Sci. Instr.* **36**, 267–270.

Furon, R. (1963). *The Problem of Water: a World Study*. Faber and Faber, London.

Gamble (1887). Quoted in Report of the Desert Encroachment Committee. Govt Printer, Pretoria, 1951.

Gear, D. (1968). Luano Catchment Expts. Agric. Res. Council of Central Africa. Ann. Rept 1967, p. 140.

Geiger, R. (1950). *The Climate Near the Ground.* Harvard University Press.

Glendening, G. E., Pase, C. P. and Ingebo, P. (1961). Proc. 5th Ann. Arizona Watershed Congress. US Forest Service. Tempe, Arizona.

Glover, P. E., Glover, J. and Gwynne, M. D. (1962). Light rainfall and plant survival in E. Africa. Dry grassland vegetation. *J. Ecol.* **50,** 199.

Glover, J. and Gwynne, M. D. (1962). Light rainfall and plant survival in E. Africa. I. Maize. *J. Ecol.* **50,** 111–118.

– and McCulloch, J. S. G. (1958). The empirical relationship between solar radiation and hours of bright sunshine in the high altitude tropics. *Quart. J. Roy. Met. Soc.* **84,** 56–60.

– , Robinson, P. and Henderson, J. P. (1954). Provisional maps of the reliability of annual rainfall in E. Africa. *Quart. J. Roy. Met. Soc.* **80,** 602–609.

Glymph, L. M. and Holtan, N. H. (1969). Land treatment in agricultural watershed hydrology research. *Water Resources Symp. No. 2,* 44–68. University of Texas, Austin.

Goode, J. E. (1968). The measurement of sap tension in the petioles of apple, raspberry and blackcurrant leaves. *J. Hort. Sci.* **43,** 231–233.

Grimes, B. H. and Hubbard, J. (1972). Aerial survey of natural resources. *Endeavour, London,* **31,** 130–134.

Gunn, D. L., Kirk, R. L. and Waterhouse, J. A. H. (1945). An improved radiation integrator for biological use. *J. Exp. Biol.* **22,** 1–7.

Hadley, R. F. and Lusby, G. C. (1967). Runoff and hillslope erosion from a high-intensity thunderstorm. *Water Resources Res.* **3,** 139–143.

– , McQueen, I. S. *et al.* (1961). Hydrologic effects of water spreading in Box Creek Basin, Wyoming. US Geol. Survey. Water Supply Paper 1532A. US Govt Printing Office. Washington, DC.

Hagan, R. M., Haise, H. R. and Edminster, T. W. (eds.) (1967). *Irrigation of Agricultural Lands.* Amer. Soc. Agron. Madison, Wisconsin.

Hamilton, E. L. (1954). Rainfall on rugged terrain. USDA Tech. Bull. No. 1096.

Hammond, R. B. (1967). Desalted water for agriculture. *Proc. Int. Conf. Water for Peace.* **2,** 184. Washington, DC.

Hanks, R. J., Allen, L. H. and Gardner, H. R. (1971). Advection and evapotranspiration of wide-row sorghum in the Central Great Plains. *Agron. J.* **63**, 520–527.

Harbeck, G. E., Jr and Koberg, G. E. (1959). A method of evaluating the effect of a monomolecular film in suppressing reservoir evaporation. *J. Geophys. Res.* **64**, 89–94.

Harrold, L. L., Brakensiek, D. L., McGuinness, J. L., Amerman, C. R. and Dreibelbis, F. R. (1962). Influence of land use and treatment on the hydrology of small watersheds at Coshocton, Ohio, 1938–1957. USDA Tech. Bull. No. 1256.

– and Driebelbis, F. R. (1958). Evaluation of agricultural hydrology by monolith lysimeters. USDA Tech. Bull. No. 1179.

– , Peters, D. B., Driebelbis, F. O. and McGuinness, J. L. (1959). Transpiration evaluation of corn grown on a plastic-covered lysimeter. *Proc. Soil Sci. Soc. Amer.* **23**, 174–178.

Hartman, A. M., Ree, W. O., Schoof, R. R. and Blanchard, B. J. (1967). Hydrological influences of a flood control programme. *Proc. Amer. Soc. Civ. Eng.* **93**, HY3, 17–25.

Hawkins, J. C. (1959). The NIAE in East Africa. *Emp. Cott. Grow. Rev.* **36**, 35.

Henrici, M. (1943). Transpiration of large Karroo bushes. *S. Afr. J. Sci.* **39**, 117.

Hewlett, J. D. and Douglass, J. E. (1968). Blending forest uses. USDA Forest Serv. Research Paper SE 37.

– and Hibbert, A. R. (1963). Moisture and energy conditions within a sloping soil mass during drainage. *J. Geophys. Res.* **68**, 1081–1087.

Hibbert, A. R. (1967), Forest treatment effects on water yield. *Proc. Int. Symp. For. Hydrol.* 527–543. Penn. State University (Pergamon Press).

Hidore, J. J. (1963). Length of record and reliability of runoff means. *J. Hydrol.* **1**, 344–354.

Hindson, L. L. and Wurzel, P. (1963). Groundwater in the Sabi Valley alluvial plain. *Rhod. J. Agric. Res.* **1**, 99–106.

Hodges, C. N. and Hodge, C. O. (1971). An integrated system for providing power, water and food for desert coasts. *Hort. Sci.* **6**, 30–33.

Holtan, H. N., England, C. B. *et al.* (1968). Moisture tension data for selected sites on experimental watersheds. USDA ARS Rept 41–44.

– and Lopez, N. C. (1971). USDAHL-70 model of watershed hydrology. USDA Tech. Bull. No. 1435.

Holtzman, B. (1937). Sources of moisture for precipitation in the United States. USDA Tech. Bull. No. 589.

Hoover, M. D. (1944). Effect of removal of forest vegetation upon water yields. *Trans. Amer. Geophys. Un.* **6**, 969–975.

(1953). Interception of rainfall in a young Loblolly Pine plantation. US Forest Service. SE For. Exp. Sta. Paper 21.

– and Leaf, C. F. (1967). Process and significance of interception in Colorado subalpine forest. *Proc. Int. Symp. For. Hydrol.* 212–222. Penn. State University (Pergamon Press).

Hori, T. (ed.) (1953). *Studies on Fogs in Relation to Fog-Preventing Forest.* Inst. Low Temperature Science, Sapporo, Hokkaido, Japan.

Horton, R. E. (1919a). Measurement of rainfall and snow. *Monthly Weather Rev.* **47**, 294–295.

(1919b). Rainfall interception. *Monthly Weather Rev.* **47**, 603–623.

(1932). Drainage basin characteristics. *Trans. Amer. Geophys. Un.* **13**, 350–361.

House, G. J., Rider, N. E. and Tugwell, C. P. (1960). A surface energy-balance computer. *Quart. J. Roy. Met. Soc.* **86**, 215–231.

Howe, G. M., Slaymaker, H. O. and Harding, D. M. (1966). Flood hazard in mid-Wales. *Nature* **212**, 584–585.

Howells, G. P. (1972). The estuary of the Hudson River, USA. *Proc. Roy. Soc. Lond. B.* **180**, 521–534.

Hudson, J. P. (1963). Variations in evaporation rates in Gezeira cotton fields. *Emp. Cott. Grow. Rev.* **40**, 253–261.

(1964). Evaporation under hot, dry conditions. *Emp. Cott. Grow. Rev.* **41**, 241–254.

(1965). Evaporation from Lucerne under advective conditions in the Sudan. *Exp. Agric.* **1**, 23–31.

Hudson, Sir William (1971). The Snowy Mountains Hydroelectric and Irrigation Scheme (Australia) *Proc. Roy. Soc. Lond. A.* **326**, 23–37.

Hursh, C. R. (1943). Water storage limitations in forest soil profiles. *Proc. Soil Sci. Soc. Amer.* **8**, 412–414.

– and Hoover, M. D. (1941). Soil profile characteristics pertinent to hydrologic studies in the Southern Appalachians. *Proc. Soil Sci. Soc. Amer.* **6**, 414–422.

– , Hoover, M. D. and Fletcher, P. W. (1942). Studies in the balanced water-economy of experimental drainage areas. *Trans. Amer. Geophys. Un.* **24**, 594–606.

225

– and Pereira, H. C. (1953). Field moisture balance in the Shimba Hills. *E. Afr. Agric. J.* **18**, 139.

International Institute of Tropical Agriculture. Annual Report for 1970–1. Ibadan, Nigeria.

Ivanov, K. E. (1953). *The Hydrology of Swamps.* Hydrometeorological Publishing House, State Hydrological Institute, Leningrad (available in Russian and Polish only).

Ivanov, K. E. (1963). *Hydrological Computations for the Drainage of Swamps.* Hydrometeorological Publishing House, Leningrad.

Jacobsen, T. (1958). *Salinity and Irrigation: Agriculture in Antiquity.* Dyala Basin Archaeological Report 1957–8. Ministry of Irrigation, Baghdad.

Jatindra Singh, Gajinder Singh and Rattan Singh (1967). Behaviour of groundwater in the Ghakra Canal Area. *Int. Comm. Irr. and Dr. Ann. Bull. 1967,* New Delhi.

Jensen, M. H. and Teran, R. M. A. (1971). Use of controlled environment for vegetation production in desert regions. *Hort. Sci.* **6**, 33–36.

Johnson, E. A. (1952). The effect of farm woodland grazing on watershed values in the Southern Appalachian Mountains. *J. Forestry* **50**, 109–113.

Jones, Graham (1971). *The Role of Science and Technology in Developing Countries.* Oxford University Press.

Kaul, R. N. and Thalen, D. C. P. (1971). Range ecology at the Institute for Applied Research in Natural Resources, Iraq. *Nature and Res.* (UNESCO) **7**, 2–15.

Kautz, H. M. (1955). The story of Sandstone Creek Watershed. USDA Yearbook for 1955, 210–218. Washington, DC.

Kerfoot, O. (1962). The vegetation of the Atumatak catchments. *E. Afr. Agric. and For. J.* **27** (Special Issue), 55–58.

  (1968). Mist precipitation on vegetation. *For. Abstr.* **29**, 8–20.

Kerrich, J. E. (1950). *An Experimental Introduction to the Theory of Probability.* University of Witwatersrand, Johannesburg.

Khosla, A. N. (1949). *J. Central Bd Irrig. Simla* **6**, 410–422.

King, H. W. (1939) *Handbook of Hydraulics.* McGraw-Hill, New York and London.

Kirby, E. S. (1972). Environmental spoilage in the USSR (quoting from *Literaturnaya Gazeta,* 17 June 1970). *New Scientist* **53**, 28–29.

Kittredge, J. (1948). Interception and stemflow. *Forest Influences,* Chap. XI, 99–114. McGraw-Hill, New York and London.

Knowles, T. and Scurlow, G. (1968). The Wagga experiments in pasture hydrology. File Report SCS of NSW.

Kohler, M. A., Nordensen, T. J. and Fox, W. E. (1955). Evaporation from pans and lakes. US Weather Bureau Tech. Paper 37, Washington, DC.

Kovner, J. L. (1956). Evapotranspiration and water yields following forest cutting and natural regrowth. *Proc. Soc. Amer. For.* 106–110.

– and Evans, T. C. (1954). A method for determining the minimum duration of watershed experiments. *Trans. Amer. Geophys. Un.* 35, 608–612.

Krestovsky, O. I. and Federov, S. F. (1970). Study of water balance elements of forest and field watersheds. *Symp. on World Water Balance.* IASH–UNESCO 2, 445–451.

Lake Hefner (1958). Evaporation reduction investigation. Report by the Collaborators. Oklahoma City: US Weather Bureau; US Geol. Survey; US Bureau of Reclamation.

Langbein, W. B. (1962). Surface water, including sedimentation. *Proc. UNESCO Symp. on Problems of the Arid Zone,* 3–22. Paris.

Langford, T. E. and Aston, R. J. (1972). The ecology of some British rivers in relation to warm water discharges from power stations. *Proc. Roy. Soc. Lond. B.* 180, 407–420.

Laycock, D. H. (1964). An empirical correlation between weather and yearly tea yields in Malawi. *Trop. Agric. (Trin.)* 41, 277–291.

– and Wood, R. A. (1963). Some observations on soil moisture use under tea in Nyasaland. *Trop. Agric. (Trin.)* 40, 35–48.

Layzell, D. (1968). Construction and use of the Line Level. Soil Conservation Service, Kenya Dept of Agric., Nairobi (Stencilled).

Lea, J. D., Ofield, R. J. and Passmore, R. G. (1960). Field observation trials of new NIAE cultivation equipment. *E. Afr. Agric. J.* 25, 220.

Ledger, H. P. (1965). The body composition of East African ruminants. *J. Agric. Sci. (Camb.)* 65, 261.

Lee, R. (1964). Potential insolation as a topoclimatic characteristic of drainage basins. *Bull. Int. Ass. Sci. Hydrol.* 9, 27–41.

(1966). Effects of tent-type enclosures on the micro-climate and vaporisation of plant cover. *Oecologia Plantarum* 1, 301–326.

– and Baumgartner, A. (1966). The topography and insolation climate of a mountainous forest area. *Forest Sci.* 12, 258–267.

Lemon, E. R., Glaser, A. H. and Satterwhite, L. E. (1957). Some aspects of the relationship of soil plant and meteorological factors to evapotranspiration. *Proc. Soil Sci. Soc. Amer.* **21**, 464–468.

Lewis, W. K. (1957). Investigation of rainfall, runoff and yield on the Alwen and Brennig catchments. *Proc. Inst. Civ. Engrs* (paper 6193).

Leyton, L., Reynolds, E. R. C. and Thompson, F. B. (1967). Rainfall interception in forest and moorland. *Proc. Int. Symp. For. Hydrol.*, 163–178. Penn. State University (Pergamon Press).

Lightfoot, L. C., Smith, S. T. and Malcolm, C. F. (1964). Salt Land Survey, 1962. *J. Agric. W. Australia* **13**, 3–10.

Linacre, E. T. (1963). Determining evapotranspiration rates. *J. Aust. Inst. Agric. Sci.* **29**, 165–177.

– , Hicks, B. B., Sainty, G. R. and Grange, G. (1970). The evaporation from a swamp. *Agric. Met.* **7**, 375–396.

Locke, G. M. L. (1970). *Census of Woodlands 1965–67: A report on Britain's forest resources.* HMSO, London.

Loeb, S. and Sourirajan, S. (1961). Sea water demineralisation by means of a semi-permeable membrane. Dept of Engineering, University of California, Los Angeles. Report 60.

Lull, H. W. and Orr, H. K. (1950). Induced snow drifting for water storage. *J. Forestry* **48**, 179–181.

Luther, H. and Rzoska, J. (1971). *Project Aqua, a source book of inland waters proposed for conservation.* IBP Handbook No. 21. Blackwell, Oxford.

McCulloch, J. S. G. (1962). Assessment of the main components of the hydrological cycle. *E. Afr. Agric. and For. J.* (Special Issue) **27**, 9.

– and Strangeways, I. C. (1967). Instrumentation in hydrology. *Proc. Int. Conf. Water for Peace* **4**, 349. Washington, DC.

– and Wangati, F. (1967). Notes on the use of the Gunn–Bellani Radiometer. *Agric. Met.* **4**, 63–70.

McDonald, J. E. (1962). The evaporation precipitation fallacy. *Weather* **17**, 168.

McIlroy, I. C. (1971). An instrument for continuous recording of natural evaporation. *Agric. Met.* **9**, 93–100.

– and Angus, D. E. (1963). The Aspendale multiple weighed lysimeter installation. CSIRO Div. Met. Physics Tech. Paper 14.

(1964). Grass, water and soil evaporation at Aspendale. *Agric. Met.* **1**, 204–224.

– and Sumner, C. J. (1961). A sensitive high-capacity balance for continuous automatic weighing in the field. *J. Agric. Eng. Res.* **6,** 252–258.

McNaughton, G. (1962). Aspects of water pollution control. WHO Public Health Papers No. 13.

Mairhofer, J. (1967). Groundwater flow and direction measurements by means of radio-isotopes in a single borehole. *Amer. Geophys. Un. Geophys. Monograph* **11,** 119–129.

Malherbe, H. L. (1968). Report of the Interdepartmental Committee of Investigation into Afforestation and Water Supplies in South Africa. Dept of Forestry. Govt Printer, Pretoria.

Martinelli, M., Jr (1967). Possibilities of snowpack management in Alpine areas. *Proc. Int. Symp. For. Hydrol.,* 225–231. Penn. State University (Pergamon Press).

Matsui, Z. (1956). The ecological features in foggy districts. For. Exp. Sta. Hokkaido, Japan, Report No. 205.

Meig, P. (1966). The geography of coastal deserts. UNESCO Arid Zone Research Report No. 26.

Midgley, D. C. (1963). Application of modern planning concepts to the development of the Orange River. *S. Afr. J. Sci.* **59,** 455–460.

Millar, A. A., Jensen, R. E., Bauer, A. and Norum, E. B. (1971). Influences of atmospheric and soil environmental parameters on the diurnal fluctuations of leaf water status of barley. *Agric. Met.* **8,** 92–105.

Miller, D. H. (1964). Interception processes during snowstorms. US For. Res. Serv. Paper PSW 18. Berkeley, California.

(1967). Sources of energy for thermodynamically caused transport of intercepted snow from forest crowns. *Proc. Int. Symp. For. Hydrol.* 201–212. Penn. State University (Pergamon Press).

Miller, R. F., McQueen, I. S., Branson, F. A., Shown, L. M. and Buller, W. (1969). An evaluation of range floodwater spreaders. *J. Range Management* **22,** 246–257.

Molchanov, A. A. (1960). The hydrological role of forests. Acad. of Sci. of USSR, Inst. of Forestry, Moscow (Translated by A. Gourevitch, Israel, 1963).

Monteith, J. L. (1957). Dew. *Quart. J. Roy. Met. Soc.* **83,** 322.

(1959a). Solarimeter for field use. *J. Sci. Instr.* **36,** 341–346.

(1959b). The reflection of short wave radiation by vegetation. *Quart. J. Roy. Met. Soc.* **85,** 366.

– and Szeicz, G. (1960). The performance of a Gunn–Bellani radiation integrator. *Quart. J. Roy. Met. Soc.* **86,** 91–94.

Morse, R. N. (1967). Solar distillation in Australia. *Proc. Int. Conf. Water for Peace.* **2**, 83. Washington, DC.

Murray, C. E. and Reeves, E. B. (1972) Estimated use of water in the US in 1970. US Geol. Survey Circ. 676.

Mustonen, S. E. and McGuinness, J. L. (1968). Estimating evapotranspiration in a humid region. Tech. Bull. 1389. ARS, USDA.

Myers, L. E. (1967). New water supplies from precipitation harvesting. *Proc. Int. Conf. Water for Peace* **2**, 631. Washington, DC.

Nace, R. L. (1970). World hydrology: status and prospects. *Symp. on World Water Balance.* IASH–UNESCO **1**, 41–48.

Nagel, J. F. (1956). Fog precipitation on Table Mountain. *Quart. J. Roy. Met. Soc.* **82**, 452–460.

Nanni, U. W. (1972). Water use by riparian vegetation at Cathedral Peak. *S. Afr. For. J.* (In press).

Natural Environment Research Council (1970). *Hydrological Research in the United Kingdom (1965–1970).* London.

Nature. (Editorial) (1953). River floods in Great Britain. **172**, 263–264.
   (1972). Ducks, drakes and barrages. **235**, 423–424.

Némec, J., Pasák, V. and Zelený, V. (1967). Forest hydrology research in Czechoslovakia. *Proc. Inst. Symp. For. Hydrol.* 31–34. Penn. State University (Pergamon Press).

Neumann, J. (1955). On the water balance of Lake Huleh and the Huleh Swamps. *Israel Exploration J.* **5**, 49–58.

Olivier, H. (1961). *Irrigation and Climate.* Edward Arnold, London.
   (1962). *Vaal: Water Balance Factors in the Development of Natural Resources.* Pub. by Vaal R. Catchment Ass., Johannesburg.

Orvig, S. (1970). The hydrological cycle of Greenland and Antarctica. *Symp. on World Water Balance.* IASH–UNESCO **1**, 1–9.

Packer, P. E. (1967). Forest treatment effects on water quality. *Proc. Int. Symp. For. Hydrol.* 687–699. Penn. State University (Pergamon Press).

– and Laycock, W. A. (1969). Watershed management in the United States; concepts and principles. *Lincoln Papers in Water Resources* **8**, 1–22.

Parsons, J. J. (1960). Fog drip from coastal stratus. *Weather* **15**, 58.

Pasquill, F. (1949). Some estimates of the amount and diurnal variation of evaporation in fair spring weather. *Quart. J. Roy. Met. Soc.* **75**, 249–256.

Patrick, J. H., Douglass, J. E. and Hewlett, J. D. (1965). Soil moisture absorption by mountain and piedmont forests. *Soil Sci. Soc. Amer. Proc.* **29**, 303–308.

Pavari, A. (1937). The influence of the Mediterranean forests on climate. *L'Alpe* **24**, 42–57. (*US Forest Service trans. No. 319*).

Penman, H. L. (1948). Natural evaporation from open water, bare soil and grass. *Proc. Roy. Soc. Lond. A.* **193**, 120–145.

(1950*a*). Evaporation over the British Isles. *Quart. J. Roy. Met. Soc.* **76**, 372–383.

(1950*b*). The water balance of the Stour catchment area. *J. Inst. Water Engrs* **4**, 457–469.

(1958). Notes on the water balance of the Sperbelgraben and Rappengraben. *Mitteilungen der Schweizerischen Anstalt fur das forsliche Versuchwesen.* Bd 35. HI.

(1962). Woburn irrigation 1951–9. *J. Agric. Sci. (Camb.)* **58**, 343–348, 349–364, 365–379.

(1963). *Vegetation and Hydrology.* Comm. Bur. Soils Tech. Comm. 53.

(1970). Woburn irrigation 1960–8. *J. Agric. Sci. (Camb.)* **75**, 69–73, 75–88, 89–102.

– and Long, I. F. (1960). Weather in wheat: an essay in microclimatology. *Quart. J. Roy. Met. Soc.* **86**, 16–50.

Pereira, H. C. (1952). Interception of rainfall by Cypress plantations. *E. Afr. Agric. J.* **18**, 1–4.

(1957). Field measurements of water use for irrigation control in Kenya Coffee. *J. Agric. Sci.* **49**, 459–467.

(1958). The limited utility of floodwater in the Perkerra River Irrigation Scheme in Kenya. *E. Afr. Agric. and For. J.* **23**, 246–253.

(1959*a*). Practical field instruments for estimation of radiation and of evaporation. *Quart. J. Roy. Met. Soc.* **85**, 253–261.

(1959*b*). A physical basis for land-use policy in tropical catchment areas. *Nature* **184**, 1768–1771.

(1964). The ranching of wild game in Africa. *Geogr. Mag.* **37**, 462–472.

(1966). Effects of land-use on the water and energy budgets of tropical watersheds. *Proc. Int. Symp. For. Hydrol.* 435–450. Penn. State University (Pergamon Press).

(1969). Influence of man on the hydrological cycle: Guide to policies for the safe development of land and water resources. FAO Working Party Report to IHD Conference, UNESCO, Paris. (Reprinted 1972 in *Status and Trends of Research in Hydrology 1965–74.* UNESCO, Paris.)

– and Beckley, V. R. S. (1952). Grass establishment on an eroded soil in a semi-arid African reserve. *Emp. J. Exp. Agric.* **21**, 1–14.

– and Hosegood, P. H. (1962). Comparative water use of softwood plantations and bamboo forest. *J. Soil Sci.* **13**, 299–314.

–, – and Dagg, M. (1967). Effects of tied ridges, terraces and grass leys on a lateritic soil in Kenya. *Exp. Agric.* **3**, 89–98.

–, – and Thomas, D. B. (1961). Productivity of tropical semi-arid thornscrub country under intensive management. *Emp. J. Exp. Agric.* **29**, 269–286.

– and McCulloch, J. S. G. (1960). The energy balance of tropical land surfaces. *Tropical Met. in Africa.* E. Afr. Met. Dept, Nairobi.

–, –, Dagg, M., Kerfoot, O., Hosegood, P. H. and Pratt, M. A. C. (1962). Hydrological effects of changes in land-use in some East African catchment areas. *E. Afr. Agric. and For. J.* **27** (Special Issue), 42–75.

–, Wood, R. A., Brzostowski, H. W. and Hosegood, P. H. (1958). Water conservation by fallowing in semi-arid tropical East Africa. *Emp. J. Exp. Agric.* **26**, 213–228.

Peterson, H. V. and Branson, F. A. (1962). Effects of land treatments on erosion and vegetation on rangelands in Arizona and New Mexico. *J. Range Management* **15**, 220–226.

Plantie, L. (1961). Technique Française Algérienne des banquettes de défense et de restauration des sols. Service des Eaux et Forêts d'Oran.

Plummer, A. P., Christensen, D. R. and Monsen, S. B. (1968). *Restoring Big Game Range in Utah.* Pub. No. 68–3, Utah Dept of Natural Resources.

Poland, J. F. (1960). Land subsidence in the San Joaquin Valley, California, and its effects on estimates of groundwater resources. Int. Ass. Sci. Hydrol. Pub. No. 52, 324–335.

Pratt, M. A. C. (1962). Relationship of runoff to rainfall. *E. Afr. Agric. and For. J.* **27** (Special Issue), 73–75.

Pruitt, W. O. and McMillan, W. D. (1962). Chapter V: *Investigation of Energy and Mass Transfers Near the Ground.* University of California: Davis College.

Rakhmanov, V. V. (1962). *The Role of Forests in Water Conservation.* Hydrometeorological Service, Moscow. (Translated by Israel Programme of Sci. Trans., Jerusalem.)

Ramey, J. T. (1967). Policy considerations in desalting and energy development and use. *Proc. Int. Conf. Water for Peace* **2**, 1. Washington, DC.

Rampacek, C. (1968). *Mineral Waste Utilisation.* Symposium: US Bureau of Mines.

Rawitscher, F. and Rawitscher, E. (1949). Inadequacy of potometry for measuring plant transpiration. *Nature* **163**, 68–69.

Reifenberg, A. (1955). *The Struggle Between the Desert and the Sown.* Hebrew University, Jerusalem.

Reifsnyder, W. E. and Lull, H. W. (1965). *Radiant Energy in Relation to Forests.* USDA Forest Service, Tech. Bull. No. 1344.

Reinhart, K. G. (1958). Calibration of five small forested watersheds. *Trans. Amer. Geophys. Un.* **38**, 933–936.

Reynolds, E. R. C. and Leyton, L. (1967). Research data for forest policy: the purpose, methods and progress of forest hydrology. *Proc. 9th Brit. Comm. For. Conf.* Pub. by Comm. For. Inst., Oxford.

Ricca, V. T., Simmons, P. W., McGuinness, J. L. and Taiganides, E. P. (1970). Influence of land-use on runoff from agricultural watersheds. *Agric. Eng.* **53**, 187–190.

Rijks, D. A. (1969). Evaporation from a papyrus swamp. *Quart. J. Roy. Met. Soc.* **95**, 643–649.

(1971). Water use by irrigated cotton in the Sudan; (3) Bowen ratios and advective energy. *J. Appl. Ecol.* **8**, 643–663.

Robinson, A. R. (1971). Sediment: our greatest pollutant. *Agric. Eng.* **53**, 406–409.

Rockney, V. D. (1960). The application of weather radar to hydrological problems. *Tropical Meteorology in Africa,* 366–377. E. Afr. Met. Dept, Nairobi.

Rocky Mountain For. and Range Exp. Sta. Ann. Report (1965). Fort Collins, Colorado.

Rodda, J. C. (1971). Progress at Plynlimon. Brit. Ass. Adv. Sci. Meeting, Swansea. Inst. Hydrol., Wallingford.

Rodier, J. (1963). *Bibliography of African Hydrology.* UNESCO, Paris.

Rosenzweig, D. (1969). Study of differences in effects of forests and other vegetative covers on water yield. Research report No. 27, Soil Cons. and Drainage Div. Hakirya, Tel Aviv.

Rothacher, J. (1963). Net precipitation under a Douglas-fir forest. *Forest Sci.* **9**, 423–429.

Rowe, P. B. (1941). Hydrology of the Sierra Nevada foothills. *Trans. Amer. Geophys. Un.* **1**, 90–100.

(1948). Influence of woodland chaparral on water and soil in California. Report Calif. Div. For.; US For. Serv.

(1963). Streamflow increases after removing woodland riparian vegetation from a southern Californian watershed. *J. Forestry* **61**, 365–370.

233

Rubcov, M. V. and Dracikov, J. K. (1970). Delimitation of protective forestry along rivers. *Lesn. Hoz.* **11**, 67–69 (in Russian). See also *For. Abstr.* **32**, 3677.

Russell, E. W. (1973). *Soil Conditions and Plant Growth* (11th Ed.) Longman, London.

Russell, Sir Frederick, and Gilson, H. C. (1972). A discussion of fresh water and estuarine studies of the effects of industry. *Proc. Roy. Soc. Lond. B.* **180**, 363–536.

Rutskovskaia, V. A. (1959). Volume changes in river flow to the Caspian Sea due to man's economic activities. *Caspian Problems,* Oceanic Commission, USSR, SA V.

Rutter, A. J. (1966a). Studies of the water relations of *Pinus sylvestris* in plantation conditions (IV). *J. Appl. Ecol.* **3**, 393–405.

(1966b). An analysis of evaporation from a stand of Scots Pine. *Proc. Int. Symp. For. Hydrol.* 403–417. Penn. State University (Pergamon Press).

(1967). Studies of the water relations of *Pinus sylvestris* in plantation conditions (VI) *J. Appl. Ecol.* **4**, 73–81.

– , Kershaw, K. A., Robins, P. C. and Morton, A. J. (1971). A predictive model of rainfall interception in forests. I. Derivation of model from observations in a plantation of Corsican pine. *Agric. Met.* **9**, 367–384.

Rycroft, H. B. (1955). The effects of riparian vegetation on water loss from an irrigation furrow at Jonkershoek. *J. S. Afr. For. Ass.* No. 26.

Saccardy, L. (1959). Nécessité de la lutte contre les erosions. Bull. Tech. de l'Inform. des Ingénieurs des Services Agricoles. No. 142 de la DRS d'Oran, Algiers.

Sanderson, M. (1950). An experiment to measure potential evapo-transpiration. *Can. J. Res.* (C) **26**, 445–454.

Satterlund, D. R. and Eschner, A. R. (1965). The surface geometry of a closed conifer forest in relation to losses of intercepted snow. US For. Serv. Res. Paper NE 34 (NE For. Exp. Stn, Upper Derby, Pennsylvania.)

– and Haupt, H. F. (1970). The disposition of snow caught by conifer crowns. *Water Resources Research* **6**, 649–652.

Sauer, S. P. and Masch, F. D. (1968). Effects of small structures on water yield in Texas. *Proc. Conf. on Watershed Changes and Streamflow.* University of Texas, Austin.

Savory, C. R. (1965). Game utilisation in Rhodesia. *Zoologica Africana* **1**, 321–337.

Schachori, A. Y. and Michaeli, A. (1965). Water yields of forest, maquis and grass covers in semi-arid regions: a literature review. *Proc. Montpellier Symp. Method Plant Eco-Phys.*, 467–478. UNESCO, Paris.

– , Stanhill, G. and Michaeli, A. (1965). Integrated research approach to study of effects of different cover types. *Proc. Montpellier Symp. Method Plant Eco-Phys.*, 479–487. UNESCO, Paris.

Schick, A. P. (1968). The inflow of fluvial sediment into Lake Kinnereth. 2nd Prog. Report, Dept of Geography, The Hebrew University, Jerusalem.

Scholander, P. F., Hammol, H. T., Hemmengson, E. A. and Bradstreet, E. D. (1965). Sap pressure in vascular plants. *Science* **148**, 339–346.

Shanan, L., Tadmore, N. H., Evanari, M. and Reiniger, P. (1970). Runoff farming in the desert (III). *Agron. J.* **62**, 445–449.

Shantz, H. L. (1954). The place of grasslands in the earth's cover of vegetation. *Ecology* **35**, 143–145.

Sharp, A. L., Gibbs, A. E. and Owen, W. J. (1966). Development of a procedure for estimating the effects of land and watershed treatment on streamflow. USDA Tech. Bull. No. 1352. Washington, DC.

Smith, E. J. (1971). Cloud seeding experiments in Australia. *Proc. 5th Berkeley Symp. Math. Stat. and Probability*. University of California Press.

Smith, L. P. (1967). Potential transpiration. MAFF Tech. Bull. 16, HMSO, London.

Soil Conservation Authority of Victoria (1966). Eppalock Catchment Project. Progress Report. SCA, Kew, Victoria, Australia.

Soumi, V. E. and Tanner, C. B. (1958). Evapotranspiration estimates from heat-budget measurements over a field crop. *Trans. Amer. Geophys. Un.* **39**, 298–304.

Sourirajan, S. (1970). *Reverse Osmosis*. Logos Press, London.

Stanhill, G. (1965a). Four methods of estimating solar radiation. *Proc. Montpellier Symp. Method Plant Eco-Phys.*, 55–60. UNESCO, Paris.

(1965b). The concept of potential evapotranspiration in arid zone agriculture. *Proc. Montpellier Symp. Method Plant Eco-Phys.*, 109–116. UNESCO, Paris.

– , Hofstede, G. J. and Kalma, J. D. (1966). Radiation balance of natural and agricultural vegetation. *Quart. J. Roy. Met. Soc.* **92**, 128–140.

Stephens, J. C. (1956). Subsidence of organic soils in the Florida Everglades. *Proc. Soil Sci. Soc. Amer.* **20**, 77–80.

Stern, W. R. and Fitzpatrick, E. A. (1965). Calculated and observed evaporation in a dry monsoon environment. *J. Hydrol.* **3**, 297–311.

Stewart, J. B. (1971). The albedo of a pine forest. *Quart. J. Roy. Met. Soc.* **97**, 561–564.

– and Oliver, S. A. (1970). Evaporation from forests. *Aberystwyth Symposia on Agric. Met.* No. 13 (G), 1–12.

Stone, E. C., Schachori, A. Y. and Stanley, R. G. (1956). Water absorption by needles of Ponderosa pine and its internal redistribution. *Plant Physiol.* **31**, 120.

Stoyer, R. L. (1967). The development of total use water management at Santee, California. *Proc. Int. Conf. Water for Peace* **2**, 478. Washington, DC.

Sumner, C. J. (1963). Unattended long-period evaporation recorder. *Quart. J. Roy. Met. Soc.* **89**, 414–417.

Sutcliffe, R. C. (1956). Water balance and the general circulation of the atmosphere. *Quart. J. Roy. Met. Soc.* **82**, 385–395.

Sutton, O. G. (1953). *Micrometeorology.* McGraw-Hill, New York and London.

Swift, L. W., Pereira, H. C., Talbot, L. M. and Beaton, W. G. (1963). Wildlife development in the savanna lands of East and Central Africa. Report No. 63–44384, UN Special Fund, New York.

– and van Bavel, C. H. M. (1961). Mountain topography and solar energy available for evapo-transpiration. *J. Geophys. Res.* **66**, 2565.

Talbot, L. M., Payne, W. J. A., Ledger, H. P., Verdcourt, L. D. and Talbot, M. H. (1965). The meat production potential of wild animals in Africa. CAB Tech. Comm. No. 16. Farnham Royal, Bucks.

Tennessee Valley Authority (1962). Reforestation and erosion control influences upon the hydrology of the Pine Tree Branch Watershed, 1941–1960. Knoxville, Tennessee.

(1963). Parker Branch Research Watershed: Project Report 1953–1962. Knoxville, Tennessee.

Thomas, H. E. and Leopold, L. B. (1964). Groundwater in North America. *Science* **143**, 1001–1006.

Thornthwaite, C. W. (1948). An approach towards a rational classification of climate. *Geog. Rev.* **38**, 55–95.

– and Holzman, B. (1939). The role of evaporation in the hydrological cycle. *Trans. Amer. Geophys. Un.* Pt IV, 680–686.

Tinker, J. (1971). The smug and silver Trent. *New Scientist* 50, 614–618.

Toebes, C., Scarf, F. and Yates, M. E. (1968). Effects of cultural changes on Makara Experimental Basin. *Bull. IASH* 13(3), 95–112.

Tomlinson, T. E. (1971). Nutrient losses from agricultural land. *Outlook* 6, 272–278.

Tothill, J. D. (Ed.) (1948). *Agriculture in the Sudan.* Oxford University Press.

Turc, L. (1961). Evaluation des besoins en eau d'irrigation, evapotranspiration potentielle. Formule climatique simplifiée et mise à jour. *Annales Agron.* 12, 13–49.

Tyson, H. N. and Weber, E. M. (1964). Groundwater management for the nation's future – computer simulations of groundwater basins. *Amer. Soc. Civil Eng. J. Hydraulics Div.* HY4: 59–77.

UN (ECOSOC) 1964. *Survey of Water Desalination in Developing Countries.* Council for Econ. and Social Affairs, New York.

US Dept of Agriculture (1957). Monthly precipitation and runoff for small agricultural watersheds in the United States. Soil and Water Cons. Res. Branch, Beltsville, Maryland.

(1958). Annual maximum flows from small agricultural watersheds in the United States. Soil and Water Cons. Res. Branch, Beltsville, Maryland.

(1960). Selected runoff events for small agricultural watersheds in the United States. Soil and Water Cons. Res. Branch, Beltsville, Maryland.

(1963). Hydrological data for experimental agricultural watersheds in the United States, 1956–59. USDA Misc. Pub. 942.

(1965). Hydrological data for experimental agricultural watersheds in the United States, 1960–61. USDA Misc. Pub. 994.

(1968). Hydrological data for experimental agricultural watersheds in the United States, 1962. USDA Misc. Pub. 1070.

(1970). Contours of change. Yearbook of the Dept of Agriculture, p. 236. Washington, DC.

US Geological Survey (1952). Water loss investigations. Vol. 1. Lake Hefner Studies. US Geol. Surv. Circ. 229.

US Nat. Acad. Sci. (1969). *Resources and Man.* Pub. No. 1703, W. H. Freeman and Co., San Francisco.

US Senate Committee (1960). Select Committee for National Water Resources: Print No. 32. Washington, DC.

237

van Bavel, C. H. M. (1961). Lysimetric measurements of evapo-transpiration rates in the Eastern United States. *Proc. Soil Sci. Soc. Amer.* **25**, 138–141.

– , Fritzchen, L. J. and Reginato, R. J. (1963). Surface energy balance in arid lands agriculture. USDA Agric. Res. Service Production Research Report No. 76. Tempe, Arizona.

– and Myers, L. E. (1962). An automatic weighing lysimeter. *Agric. Eng.* **43**, 580.

van Helmont, J. B. (1652). Complexionum atque mistionum elementalium figmentum. Ortus Medicinae, 84–90. Amsterdam. (See Russell, E. W. (1973). *Soil Conditions and Plant Growth*. Longman, London.)

Vilentchuk, L. (1967). Desalted water for Israel's agriculture. *Proc. Int. Conf. Water for Peace* **2**, 13. Washington, DC.

Voight, G. K. (1960). Distribution of rainfall under forest stands. *Forest Sci.* **10**, 277–282.

Walker, H. O. (1956). Evaporation. *J. W. Afr. Sci. Ass.* **2**, 107–121.

Wallihan, E. F. (1940). An improvement in lysimeter design. *J. Amer. Soc. Agron.* **32**, 395–404.

Wallis, J. A. N. (1963). Water use by irrigated Arabica coffee in Kenya. *J. Agric. Sci.* **60**, 381–388.

Walter, H. (1964). Productivity of vegetation in arid countries. *Abstracts 10th Int. Bot. Cong.*, 9. Edinburgh, Constable.

Ward, H. K. and Clegghorn, W. B. (1964). The effects of ring-barking trees in *Brachystegia* woodland on the yield of veld grasses. *Rho. Agric. J.* **61**, 98.

Ward, P. R. B. and Wurzel, P. (1968). The measurement of riverflow with radioactive isotopes with particular reference to the method and time of sampling. Bull. IASH **13**, 40–48.

Water Resources Board (1966). Water Supplies of South East England. Pub. No. 1. HMSO.

Watkins, L. H. (1962). The design of urban sewer systems. Rd Res. Lab. Tech. Paper No. 55. Dept Environment, Crowthorne, Berks.

Weinberg, A. G. (1969). Nuclear energy and the agro-industrial complex. *Nature* **221**, 1278.

Webster, C. C. and Wilson, P. N. (1966). *Agriculture in the Tropics*. Longman, London.

West, B. G. (1958). The soils of Iraq and their management. *Prospects of Iraq Biology* **1**, 15. Pub. by the Biological Society of Iraq, Baghdad.

Whyte, R. O. (1958). International development of grazing and fodder resources. V. India. *J. Brit. Grassland Soc.* **13**, 203–212.

Wicht, C. L. (1941). An approach to the study of rainfall interception by forest canopies. *J. S. Afr. For. Ass.* **6**, 54–70.

(1943). Determination of the effects of watershed management on mountain streams. *Trans. Amer. Geophys. Un.* **2**, 594–608.

(1967*a*). Forest hydrology research in the S. African Republic. *Proc. Int. Symp. For. Hydrol.*, 75–84. Penn. State University (Pergamon Press).

(1967*b*). Validity of conclusions from S. African multiple watershed expts. *Proc. Int. Symp. For. Hydrol.*, 749–758. Penn. State University (Pergamon Press).

Wiener, A. (1967). The role of advanced techniques of groundwater management in Israel's national water-supply system. Report T. 168. Tahal (Water-Planning for Israel), Tel Aviv.

Willatt, S. T. (1971). Model of soil water use by tea. *Agric. Met.* **8**, 341–351.

Willcocks, Sir William (1911). *Irrigation of Mesopotamia.* Spon, London.

Wilm, H. G. (1943). Efficient sampling of climatic and related environmental factors. *Trans. Amer. Geophys. Un.* **24**, 208–212.

(1949). How long should experimental watersheds be calibrated? *Trans. Amer. Geophys. Un.* **30**, 272–278.

Wisler, C. O. and Brater, E. F. (1959). *Hydrology.* John Wiley and Sons, New York.

Wolman, M. G. and Schick, A. P. (1967). Effects of construction on fluvial sediment, in urban and suburban areas of Maryland. *Water Resources Res.* **3**, 451–463.

Woodhead, T. (1970). Mapping potential evaporation for tropical East Africa. *Symp. World Water Bal.* IASH–UNESCO **1**, 232–241.

Wurzel, P. and Ward, P. R. B. (1965). A simplified method of groundwater direction measurement in a single borehole. *J. Hydrol.* **3**, 97–105.

(1967). Radio-isotope measurements of the movement of groundwater in the Sabi Valley. *Rho., Zamb. and Malawi J. Agric. Res.* **5**, 199–209.

Yates, M. E. (1971). Effects of cultural changes on Makara Experimental Basin: hydrological and agricultural production effects of two levels of grazing on improved and unimproved small catchments. *N.Z. J. Hydrol.* **10**, 59–84.

Zinke, P. J. (1967). Forest interception studies in the United States. *Proc. Int. Symp. For. Hydrol.*, 137–162. Penn. State University (Pergamon Press).

Subject Index

Adelaide, South Australia, 188
advection effects, in evaporation, 55, 67–68, 105, 122, 165
aerodynamic methods, for measuring evaporation, 68–9
Africa: semi-arid grassland in, 160; tropical forest studies in, 117–19; *see also individual states*
Agricultural Research Council, UK, 207
agriculture: effects of, on water resources, 167–70, 183; in lands with seasonal drought, (summer) 170–5, (winter) 175–6; in lands with water excess, 180–1; pollution caused by, 206–8; in semi-arid lands, 181–3; subsistence, spread of, 212; in tropics, 176–80
air photographs: for land-use records, 70; for mapping, 39
Akosombo Dam, Ghana, 7
albedo, 93, 165
alfalfa, evaporation from, 68
Algeria, 182–3
Amazon R., 5
Andes, soil erosion in, 27–9
Angostura Dam, Wyoming, 161, 163
Antarctica, ice-cap of, 4
aquifers, 11; location of, 40, 50, 51; intrusion of salt water into, 14; pumping from, 13; specific yield of, 92; *see also* groundwater
Argentina, 27–9
Arizona, 74, 76, 161, 163, 190
Ashdod, new Israeli port, 194
Aswan High Dam, Egypt, 7
atomic energy, for desalination, 17–18, 21
Atomic Energy Authority, UK, 16
Australia, 64, 143; advection effects in, 68; cloud-seeding in, 21; desalination in, 17; marshes in, 196; mist condensation in, 73; overgrazing and erosion in, 154–7; salt lands of, 76–8; *see also* Murray R., Snowy Mountains Scheme

bamboo forest, water use by, 131–8
banana plantation, transpiration by, 62
base-flow recession curves, 50, 91
Bermejo R., Andes, 27, 28–9
bilharziasis, 191–2
Birkenhead, 144

Blackwood R., West Australia, 77
Bolivia, 27, 28
Boulder Dam, USA, 7
Brazil, 27, 28
Brennig R., North Wales, 144–5
Britain, *see* United Kingdom
Buenos Aires, 28, 29
bush control, on rangeland, 142, 148, 153

California, 17, 73; aquifers in, 14, 190; fire damage in, 76; sewage treatment in, 13, 203
canals (irrigation): sedimentation in, 22; seepage from, 186, 191
Caspian Sea, 187
Cenchrus ciliaris (buffel grass), 154, 158, 159
Ceylon (Sri Lanka): irrigation in, 22–3; land-use problems in, 210
Chad Basin, 37
Chattahoochee R., Georgia, 174
chemical tracers, for measurement of streamflow, 46, 47
Cheyenne R., Wyoming, 35, 161
cities, stormwater from, 201–2; *see also* urban regions
climate, 4; dependence of transpiration on, 89–91; effect of forests on, 72–4
cloud cover, estimation of radiation from, 65, 66
cloud-seeding: for hail prevention, 21; for rain-making, 20–1
coffee plantation, water budget for, 128–9, 130
Colorado, 14
computer applications: for cartography, 70; for simulation, 12, 33, 109; for studies of change in land use, 135, of energy balance, 68, of transpiration, 60; of underground water, 53, and of water balance, 109–13
Congo Basin, 74
conifer plantations, 86, 95, 96, 99; water use by, 131–8, 139–40; in marshes, 196
contour farming, 24, 168, 169, 175, 176–180
conversion factors for units of measurement, xiv
Coshocton Soil Conservation Experiment Station, Ohio, 60, 138, 170–1, 174

cotton, 193; advection effects on, 67
Coweeta Hydrological Laboratory, Appalachian Mts, 85, 86, 87, 88, 96, 98, 146
Cynodon dactylon (Bermuda or star grass), 158

dams, 6, 7; for flood control, 33–4; *see also individual dams*
Dee estuary, Cheshire, 14
desalination of water, 15–20
deserts: coastal, 5, 17; expansion of, 79, 165; produced by overgrazing, 151
developing countries: land-use changes in, 115; need for hydrological information in, 38–9, 39–41; pollution in, 208–9
dew, 2, 5–6
Dnieper R., 73
domestic use of water, amounts of, 10
drainage: of lysimeters, 60–1; of marshlands, 194–200; ploughing for, 24, 169, 178; required for successful irrigation, 185, 186; of sphagnum-moss swamps, 197–8
drought, 160

ecology, 25, 72–4, 78, 142, 153, 199
economics: of boring for aquifers *v*. dam-building, 50, 51; of desalination, 15–20 *passim*; of irrigation, 107; of soil conservation, 24–5; of waste disposal, 203; of water-supply, 32, 33
education: in contour banking, 179–80; of public in hydrology, 7, 25
Egypt, 6, 7; *see also* Nile R.
Eichhornia crassipes (water hyacinth), 192
electrical resistance, measurement of moisture tension by, 54
electrodialysis, desalination by, 15
Eppalock catchment scheme, Victoria, 154–7
estuaries, storage of water in, 11, 14–15
Ethiopia, 177, 178, 211; water harvesting in, 193–4
eucalyptus plantations, 78
Euphrates R., 22
eutrophication, 206
evaporation, 55; average annual amount of, UK, 90; from land and from ocean, 5; measurement of, 55–8, 151; Penman equation for, 56, 64, 66–7, 68, 89–90, 122; potential rates of, 55–6; from reservoirs, 35–6; from vegetation, 58–9
evaporation pans, 55, 56–8
evapotranspiration, 55, potential, 63–7

filtration: desalination by, 16; of water in marshes, 28, 198

fires in forests, hydrological effects of, 75–6, 78, 79, 161, 163
floods, 5, 23, 69; dams to control, 33–4; effects of marsh drainage on control of, 198–9; measurement of streamflow during, 49; spreading of, on rangelands, 162–5; use and storage of water from, 193–4
Florida: desalination in, 16; Everglades in, 199–200
flumes, for measurement of streamflow, 48–9
Forest Reserves, 25, 30, 118
forests: control of stormflow by, 107–9, 125; destruction of, by fire, 75–6, 78, 79, 161, 163; effects of, on groundwater, 76–8, 91, 134, on snowmelt, 104, and on streamflow, 78–89, 180–1; effects of grazing in, 145–8; fog and mist condensation by, 72–3, 102; interception of rain by, 98–102; in protection of streamsources, 25; reflection of radiation from, 96, 97; replanting of, 27, 81–3; trapping of snow by, 73–4, 78, 103–4; tropical, experiments in, 115–19; water use by, (bamboo and conifer) 131–8, (rain forest and tea) 119–27, (scrub and conifer) 139–40
'fossil' water, 12
freezing, desalination by, 15, 16

governments: and hydrological information, 69–71, 72; and organisation of irrigation, 22–3, 185; and water resources, 25, 26–7, 36, 203, 214
grain crops, 181–2, 193
grassland, 142–3; in crop rotations, 169; in forests, 145–8; reflection of radiation from, 165–6; streamflow from drained and marshy, 144–5, and from poor and improved, 143–4, 153; tropical, regeneration of, 153–4; water use by, 158–9; *see also* rangelands
grazing: on dry grassland, 149, 160; in forests, 145–8; on marshland, 144–5; in protected parklands, 32; *see also* overgrazing
Greece, 25–6
'green revolution', 186
Greenland, ice-cap of, 4
groundwater, 4, 11–13; effect of forests on, 76–8, 91, 134, 138; measurement of, 50–3; overpumping of, 14, 190–1; raising of level of, by irrigation, 185; recharging of, 13, 51, 134, 158, 193; saline, 4, 13–14, 77–8, 188; seasonal changes in, 91–2; *see also* aquifers
Guernsey, desalination in, 15
gypsum block tensiometers, 54, 127

hail prevention, 21
Hawaii, 35, 102
hexadecanol film, to prevent evaporation, 35–6
Hudson R., New York, 202, 205
Huleh swamps, Israel, 198
hydrograph, 49
hydrological cycle, 1–4, 89

ice-caps, as reservoirs, 4
Incas, terraces of, 177, 180
India, 6, 149; irrigation in, 186; water harvesting in, 193
industry: water demand by, 10, 80; water pollution by, 10, 14, 80, 202–6
Institute of Hydrology, UK, 23, 96, 113, 118, 120, 199
international aid, 9, 39, 211, 214
International Atomic Energy Authority, 51, 52
International Biological Programme, 200
International Commission for Irrigation and Drainage, 184
International Hydrological Decade, 41, 72, 118, 153; symposium of, in New Zealand, 89, 120
International Institute of Tropical Agriculture, 181
Iraq, 6, 159, 210; irrigation in, 22, 185
irrigation, 10, 184; biological hazards of, 191–2; calculation of soil moisture balance for, 129, 131; dams for, 7; desalination of water for, 14, 17–18; government organisation of, 22–3, 185; problems of salinity and drainage in, 184–90
Israel, 62, 68, 91; cloud-seeding in, 21; desalination of water in, 15, 17; reflection by land surfaces in, 96; sewage treatment in, 203; swamp drainage in, water resources of, 13, 14, 194
Italy, 21, 95

Japan, 73, 102
Jinja Dam, Nile, 40
Johannesburg, 34
Jordan, 6
Jordan R., 13, 14, 198

Kalahari desert, 165
Kariba Dam, 7
Kenya, 40, 67, 102; forest experiments in, 117–29, 131–8; hail prevention in, 21; reflection measurements in, 95; soil conservation in, 169, 179; water budget for coffee in, 128–30; wildlife in, 31
Kenya Tea Company, 120
Kuwait, 14, 15

land use: difficulties in control of, 209–11; need for experiments on, 113–14, 140–141; recording of changes in, 69–71, 72, 76–8
leakage, from river basins, 91, 92–3; checks for, in calculations of water balance, 122–5, 136, 138
'line level' equipment, for making contour banks, 179
livestock: excreta of, 206–7; watering points for, 35, 161, 165, 166; see also grazing
logging, erosion caused by, 147–8
Los Angeles, 10
Lynmouth flood, 23
lysimeters, 55, 59–61, 111, 125

maize: advection effects on, 68; improved varieties of, 186
Malawi, 129–31
mapping: by aerial survey, 39; automated, 70
marshlands: drainage of, 194–200; effect of grazing on, 144–5; filtration of water by, 28, 198
masts, for instrumentation, 67, 95
Mesopotamia, see Iraq
metals, pollution by, 204
Mexico, desalination of water in, 16, 17, 18–19
Mississippi R., 69, 110, 174
mists, condensation onto trees of, 72–3
models: electrical, 12, 53; mathematical, 33, 52–3, 110; physical, 110
molybdenum deficiency, 155
Montana, 163, 164
Morecambe Bay, 14
Murray R., Australia, 7, 33; irrigation schemes and, 187–90
Murray River Commission, 188

Nairobi, 131
national parks, 14, 30, 199, 200
National Trust, UK, 31–2
Natural Environment Research Council, UK, 15, 23, 113
nature reserves, 199
Nebraska, 171
Negev desert, 35, 192
neutron-probe, for determining soil moisture, 53–4, 127
New Jersey, 13
New Mexico, 52, 163
New York, 10, 14, 205
New Zealand, 143
Nigeria, 181
Nile R., 6, 7, 40, 195
nitrogen, in waste water, 206
Norfolk Broads, 15

nuclear power, for desalination of water, 17–18, 21

ocean: evaporation from, 5, 57; water held by, 4
Ohio, 139, 140
oil-fuelled processes for desalination of water, 18–19
Oklahoma, 34, 36
Orange R., 34
orientation of catchment areas, 97–8
overgrazing, 79; restoration of land eroded by, 27, 149–57
oxygenation of Thames water, 203–4

Paraguay, 27, 28, 37
Parana R., 26–9, 198
Perth, Western Australia, 77–8
Peru, irrigation in, 23
Phalaris tuberosa, pasture grass, 155
phosphates, in waste water, 206
Phytolacca decandra, snail-killing compound from, 192
plants: internal water stress in, 63; protection of soil surface by, 169; roots of, *see* roots; on streambanks, 80, 105–7; transpiration by, *see* transpiration
Plato, on erosion, 25–6, 212, 214
ploughing: contour, 24, 169; downhill, 24, 169, 178; with tied ridges, 176, 177, 178
politics, and hydrology, 7, 186, 209–11
pollution: detection of, 70, 71; in developing countries, 208–9; by industry, 9, 10, 14, 80, 202–6; from modern agriculture, 206–8; by sewage, 9, 11, 202
polystyrene panels, to reduce evaporation from water surface, 36
potometers, 62–3
power stations, waste heat from, 206
protein, in grass, 160–1
pumping tests, of groundwater, 92
Punjab, 186

rabbits, contribute to overgrazing, 155
radar, measurement of rainfall by, 44
radiation: estimates of, for tropics, 65–6; measurement of, 65; measurement of reflection and of, 90, 93–8, 165–6; net, 65; *see also* solar energy
radioactive tracers: for groundwater, 12–13, 51–2; for measuring streamflow, 46
radioactivity, measurement of snowfall by, 45
radiocarbon dating, of ice, 4
rainfall, 5; and arable farming, 181; cloud-seeding for, 20–1; forest influence on,

74; intensity of, 44; interception of, 98–102; measurement of, 38, 39, 41–4, 151; recharge of groundwater by, 51, 134; reliability of, 181
rain forest, 122, 124, 180–1
rain-making, *see* cloud-seeding
rangelands, 142, 148–9, 159–61; flood-spreading on, 162–5; improvement of, 149–53, 154–7; snow-trapping on, 145; water use by improved, 158–9
reafforestation, effects of, 27, 81–3
recreational use of land, 15, 29–30, 208
reflection, *see under* radiation
resources, development or destruction of, 213
Rhine R., 202, 205–6
Rhodesia, 7, 31, 40, 46
rivers: dams controlling, 7; discharge into ocean by, 5; *see also individual rivers, and* streamflow
Road Research Laboratory, UK, 201
roads: on drainage divides, 36; erosion caused by, 209; stormwater from, 202
roots of plants: depths of, 1, (bamboo and conifers) 133, (coffee) 128, (grasses) 143, 158, (tea) 129; effect of flooding on, 193; penetration of, through soil pore space, 168
Rothamsted Experiment Station, 10
Royal Commission on Environmental Pollution, UK, 204, 207–8

Sahara desert: spreading, 165; water under, 12–13
Salvinia auriculata (water fern), 192
Saudi Arabia, 194
sealing of catchment surfaces, 35, 176
sediment accumulation: in irrigation channels, 22; in reservoirs, 13, 157, 160, 167; in rivers, 45
sediment flow, *see* soil erosion
seepage: from canals, *see* leakage; of fresh water into ocean, 5; from rivers into aquifers, 51
Seine R., 202
Severn R., 113, 199
sewage: pollution by, 9, 11, 202, 205; treatment of, 13, 16, 186, 203
simulation techniques, with computer, 12, 33, 104
snow: measurement of fall of, 38, 44–5; melting of, 104, 197; trapping of, by forests, 73–4, 78, 103–4; and on rangelands, 145
Snowy Mountains Scheme, Australia, 7, 21, 33, 75–6, 189–90
soil: air in, 168; depths of, 1
soil conservation, 23–5, 34, 157, 168–9, 170–80, 189–90

soil erosion and sediment flow: afforestation and, 83; in Andes, 26–9; from arable farming, 167, 173–5, 182–3; after fires, 75–6, 161, 163; from logging, 147–8; from overgrazing, 146, 155–7

soil moisture, 1; budgets for, 104, 124, 127–31, 132; capacity of soil for, 108, 163; deep-rooted grasses and, 158; grain crops and, 181–2; measurement of, 53–4, 127, 192; in overgrazed savannah, 152, 153; seasonal changes in, 91

soil structure, 167–8

solar energy, 1, 64; desalination by, 16–17; evaporation by, 55, 89, 101; *see also* radiation

sorghum, evaporation from, 68–9

South Africa, 24, 31, 34, 44, 102; cloud-seeding in, 21; forests and streamflow in, 78–80, 86; streambank vegetation in, 80, 105

spacecraft, photography of cloud from, 5, 38

sphagnum-moss swamps, 196–8

Sri Lanka, *see* Ceylon

State Hydrological Research Institute, Leningrad, 175, 196

storage of water, underground, 11, 13–14, 194

stormflow: arable farming and, 170–1, 173, 174; from cities and roads, 201–2; control of, 107–8, 125–7; measurement of, 151–2; records of, 49

streamflow, 90; analysis of records of, 49–50, 84–9; effects on, of change in land use, 72, of forest fires, 75–6, 163, of forests, 78–81, and of rain forest and tea plantation compared, 119, 122, 125; measurement of, 38, 40, 45–9, 84–8

strip cropping, 169

Sudan, 67, 193, 195

sulphuric acid, in soil, 195

swamps, *see* marshlands

Switzerland, 84–5

Tamarix pentandra (tamarisk), water use by, 105

Tame R., 205

Tanganyika, Groundnut Scheme in, 40

Tanzania, 5, 67, 73, 117, 165

tea plantations: measurement of transpiration in, 61; replacing rain forest, 119–25; soil moisture budget for irrigation of, 129, 131; stormflow from, 125–7

Tea Research Institute, 120

Tennessee R., 32–3, 208

Tennessee Valley Authority, 7, 32–3, 172–4, 208; reafforestation by, 81–3

tent method of estimating transpiration, 62

terracing, 177, 182

Texas, 34

Thames R., 11, 13, 38, 202, 203–4, 205

Tigris R., 22

tornados, 23

transpiration by plants, 1–2, 58–9; climatic factors in, 63–7, 89–91; discrimination between evaporation and, 100–1; measurement of, 59–63

transpiration ratio, 59

Trent R., 16, 205

Trifolium subterraneum (clover), 155

tritium in rainfall, 12, 52

Turkey, 26, 27

Uganda, 31, 40, 67, 117, 177; marshes in, 195, 196; restoration of overgrazed savannah in, 149–53

Ukraine Institute of Hydrometeorology, 175

Union of Soviet Socialist Republics (USSR): dams in, 7; measurements in, of rainfall, 42, 70, of snowfall, 44–5, of streamflow, 46, 49, 175, of transpiration, 60–1, and of water and energy balances, 95–6; shelter belts of forest in, 78; snow-trapping by forests in, 73–4, 78; sphagnum-moss swamps in, 196–8

United Kingdom (UK), 32; floods in, 23; measurements in, of evaporation, 66, and of reflection, 96; pollution in, (agricultural) 206–7, (industrial) 203–5, 207–8; water balance in, 90, 180; water use in, 10

United Nations, 9, 15, 29, 30, 36–7, 211

United States of America (USA): agricultural pollution in, 206; collection of water in, 35; control of forest grazing in, 146–7; dams in, 7, 143; flood control in, 34; land use studies in, 85–6, 111, 138–40; measurements in, of evaporation, 56–7, of rainfall, 42, 44, and of streamflow, 46, 48, 85–6, 87–8; range management in, 163; snow studies in, 103; soil conservation in, 23–4, 34, 170; water law in, 26; wildlife in, 30, 31

urban regions, water requirements of, 29, 37

Utah, 146

Vaal R., 34

Volga R., 7, 69, 73; irrigation schemes and, 186–7

Wash, the, 14

wastes: cost of disposal of, 203; water in

wastes (*cont.*)
    disposal of, 9–10, 10–11; *see also* sewage
water: amount of, in circulation, 4–6; balance of, 64, 89–91, 98; demand for, 9–11, 32–7, 201; harvesting of, 35, 192–4; management for maximum output of, 32; quality of, 70; in soil, *see* soil moisture
Water Conservation Laboratory, USA, 35
Water for Peace Congress, 9, 17
Water Pollution Research Laboratory, UK, 16, 203, 206
Water Resources Board, UK, 10, 16, 36, 113
weirs, in measurement of streamflow, 40, 45, 46–8

wells, 11, 91, 185
wheat: evaporation from, 68; improved varieties of, 186
wildlife, 14, 116, 147; on dry grassland, 149; grazing in forests, 147–8; and hydrological measurements, 42, 116; management of populations of, 30–1
wind-run, for evaporation calculations, 66, 67
World Meteorological Association, 40, 42, 56
Wye R., 113, 199
Wyoming, 161, 163

Zambia, 7, 31, 117